S0-AIF-408

Homebrew Wind Power

TJ
820
.B376

#2948 85893

Homebrew Wind Power

A HANDS-ON GUIDE TO HARNESSING THE WIND

By DAN BARTMANN AND DAN FINK

FOREWORD BY MICK SAGRILLO

Buckville Publications LLC

Masonville, CO | www.buckville.com

LIBRARY
WAUKESHA COUNTY TECHNICAL COLLEGE
800 MAIN STREET
WITHDRAWN
PEWAUKEE. WI 53072

Homebrew Wind Power: A hands-on guide to harnessing the wind

Copyright © 2009 by Dan Bartmann and Dan Fink

No part of this book may be reproduced, stored in a retrieval system or transmitted in any form, or by any means, electronic, mechanical, photocopying, recording or otherwise, without prior written notice permission of the publisher, except by a reviewer, who may quote brief passages in review.

Published by Buckville Publications LLC — PO Box 292, Masonville, CO 80541 USA

First edition 2009

ISBN 978-0-9819201-0-8

Library of Congress control number (LCCN): 2008933538

Editor: Dan Fink

Book design by Dan Fink

All photographs and drawings by the authors unless otherwise noted

Front cover photograph by Tim Fecteau

Cover design by LaVonne Ewing

Printed in Canada

By printing this book on chlorine-free, 100 percent post-consumer waste recycled paper, Buckville Publications LLC saved 36 fully grown trees, 12,968 gallons of water, 25 million BTUs of energy, 1,665 pounds of solid waste, and 3,124 pounds of greenhouse gases. These calculations are based on research by Environmental Defense and the Paper Task Force. The book was manufactured by Friesens Corporation.

CYA STATEMENT: Due to the fickle nature of wind power, Buckville Publications LLC and the authors assume no responsibility for personal injury, property damage, or loss from actions inspired by information in this book. Electrical wiring, wind turbine building, and tower construction are inherently dangerous activities, and should be regarded as such by the reader. Always consult with your local building codes and the National Electric Code before installing a renewable energy system, wind turbine, or tower. When in doubt, ask for expert advice—the recommendations in this book are no substitute for the directives of equipment manufacturers, tool manufacturers, chemical manufacturers' Material Safety Data Sheets, and federal, state and local regulatory agencies.

"Nothing is too wonderful to be true if it be consistent with the laws of nature."

— MICHAEL FARADAY

Table of contents

Foreword BY MICK SAGRILLO

A spinning wind generator has an enormous amount of emotional and eye appeal, similar to watching waves on a beach or flames in a campfire. People are mesmerized by the motion, and infatuated by the thought that simple motion caused by the wind can produce something useful—electricity. Invariably, thoughts come around to "I can build one of those!"

My initial response whenever questioned about what's involved in building a wind generator from scratch is, "Don't go there!"

"Wow, is that guy negative!" you say? I began tinkering with renewable energy in 1973, spurred on by the first oil embargo. That journey took me through almost all of the renewable energy technologies: solar hot air collectors, solar water heaters, passive solar structures, wood stoves, water pumping windmills, small-scale hydroelectric systems, photovoltaics (PV), and, of course, wind electric systems. Without question, the most difficult renewable resource to master (including, I think, casting your own PV cells) is wind electric systems. Never one to give up on a challenge, I've been seriously involved in the field of wind power for nearly three decades.

There are smart people who have spent years trying to perfect their wind generator designs, only to have them fail in the marketplace due to overly ambitious performance claims and insufficient engineering. There are dozens of patents on the books for all sorts of "whiz-bang" wind devices that never made it beyond paper and ink. Small wind generator manufacturing companies doing business in the USA today, including imported machines, barely number in the double digits. That's why I chuckle when some dreamer thinks he can design and build his own wind generator.

But, like myself, some people just can't be persuaded to try something easier, like designing their own car. This isn't to besmirch the successful manufacturers selling their creations. Far from it! They have succeeded where hundreds of others have failed. So, it can be done. Henry Ford *did* begin his business in a garage, so anything is possible. In fact, trying to design and build your own wind generator can be an incredible—albeit complex and time consuming—learning experience, taking you from mathematics and physics to fluid dynamics, material sciences, and dynamics engineering.

There is no better science project to test one's problem-solving ability than trying to build a wind generator. In fact, for those involved with high school and junior high school students, or even a college tech school, building a wind generator can be a multidisciplinary endeavor for a group of impassioned students. This doesn't mean that they'll actually build something that will work, or that their creation will stay together through the first storm that happens by. But they will certainly get a lifetime of lessons in trial-and-error problem solving.

Which takes us back around to the question, "Why would you want to take on such a project if the success rate is so miserably low, even almost non-existent?" As a hopeless tinkerer, this question makes no sense to me. My response is: "Why would you *not* want to harness the wind?" Well then, why not buy someone else's warranted product?

Let's look at some of the reasons you might want to buy an actual commercial product that is successful in the marketplace, rather than build your own wind turbine.

• The environment that these devices live and operate in is brutal. Daytime temperatures of a painted surface in the sun atop a tower in mid-summer can easily hit 160 degrees F or more, while wintertime temps can plunge to -30 degrees F or so. Temperature swings like that wreak havoc on coatings like paint and motor winding varnish, to say nothing of fasteners, composite materials, adhesives, and something as seemingly simple as tolerances between a bearing and its race, or the air gap of an alternator. Now throw in some moisture from rain, snow, or just humidity into the mix. Add a bit or a bunch of wind, your choice, but don't forget to include the bugs, dust, and debris that will tag along in a good gale. What you have is a classic materials science test bed for stressing the products you have chosen to assemble your creation with.

• If you haven't figured this out yet, Nature's continuous entropy is hard at work trying to disassemble your wind turbine and return it to her pool of resources to mine for future creations. Time is on her side, not yours. Don't forget that you will want your wind turbine to operate for oh, say, 20 years. Maybe 30. Not for just a few weeks or even a few months.

• Anyone familiar with the wind understands that it can be, and often is, cruel and unusual punishment. Maybe not most of the time, but only once in a while when it really picks up and goes ballistic. Unfortunately, you usually cannot choose what wind speeds you allow your turbine to run in—you get what's given to you and either survive or design and build a second version. Or a third. Or a fourth...

• And then there's gravity. If you haven't noticed, wind turbines are mounted on tall towers to avail themselves of the wind resource. You would no more install a wind turbine on a short tower than you would a solar collector in the shade of a tree or a hydro plant on the bank of a river. And with towers of any height comes gravity. Constance, the goddess of gravity, will continuously test her prowess against your designing abilities, forethought and desire to keep your creation upright. All too often with homebuilt wind turbines and towers, Constance wins.

So, why exactly are we gathered here? Why are we doing this? To build a wind turbine to generate a bit of electricity that could much more easily be supplied by the local utility or a PV array? Are we slow learners, or what?

Again, just look at the plethora of start-up wind companies over the last few decades and the dearth of successful products in the marketplace. Some of us are either terminally slow learners, or so stubborn that we are blind to the reality of the task at hand. But as a hopeless tinkerer, this conclusion makes no sense to me.

Let's put a positive spin on the question and just say that some of us truly love a challenge, even when the odds are so obviously stacked against us. So, you ask, why will the design in this particular book work when so many others have failed? Fair question, and give yourself a point for asking.

The design that Dan and Dan and George and Rich offer originated with Hugh Piggott of Scoraig Wind Electric in Scotland. There are very few people in the world who truly deserve the title of "genius," but Hugh is one of them, one of the very best in the area of wind technology. His musings and ruminations on the hows and whys of capturing the wind and successfully converting it into electricity have brought a deep admiration for his abilities from all of us in the small wind community. Hugh is the best of our best.

The Otherpower crew attended one of the first do-it-yourself wind generator workshops that Hugh taught in the US a number of years ago, setting them on the correct path of learning and experimentation—and learn they did. With nearly two hundred wind turbines built (and many destroyed) since then, they have taken Hugh's basic design and fine-tuned it to a point where their turbines could actually be successful commercial products, should they ever decide to go that route. These guys will save you at least several hundred years of experimentation.

The Dans, George, and Rich also exude an incredible atmosphere of enthusiasm for what they do. Undaunted by scores of failures, they get right back up and try again. And they convey this drive in the workshops that they teach. I have hosted two do-it-yourself workshops in my shop taught by these guys. I've watched over three dozen folks of all skill levels burn out drill bits, weld parts upside down and backwards, wire coils incorrectly, and chomp into blade billets like blind beavers. Unfazed by mishaps, the Otherpower crew patiently explains what went awry and brainstorm to redirect the work back on track. Folks leave not only learning the "hows," but also understanding the "whys" of what they did.

But the Otherpower crew also started out with a collection of skills that, unfortunately, not many people have anymore. Metal fabrication, welding, machining, woodworking, math and physics, and above all, thinking and problem solving. Be forewarned: If you take on building a wind turbine under the tutelage of this book, you will undoubtedly learn a lot of things you never dreamed you even needed to build a wind turbine. Like patience. And a sense of humor.

You will also shed any arrogance that might be lurking in the back of your mind as you lose your smugness towards Nature and Constance. This is hard, time consuming, and humbling—but very satisfying—work. While, to the critical eye, this turbine design may look "crude" by the standards of manufactured products, it will be very heavy-duty and quite robust. An impressive piece, indeed. If you finish reading the book, learn some new skills, stretch your abilities and persist, you will build a wind turbine, and it will work. And I have no doubt that it will supply you with decades of wind-generated electricity. It can be done!

Yes!

Mick Sagrillo

Sagrillo Power & Light

Wisconsin, USA

About Mick Sagrillo

Mick Sagrillo is the wind energy specialist for Focus on Energy, the State of Wisconsin's renewable energy benefits program. He was a founding member and current board member of the Midwest Renewable Energy Association (MREA) and RENEW Wisconsin. He has served as the president of the MREA for the past 18 years. Mick was the founder and former owner of Lake Michigan Wind & Sun for 17 years, a company that specializes in residential wind energy systems.

Mick has written numerous articles for *Home Power Magazine*, the American Wind Energy Association's *Windletter*, the American Solar Energy Society's *Solar Today Magazine*, and numerous other publications. He teaches wind energy installation courses for the MREA, Solar Energy International, and other renewable energy organizations. He also does consulting on home-sized wind systems nationally and internationally.

Photo by Dan Chiras

Preface

Winters can be tough way up here in the Northern Colorado rocky mountains at 8,200 feet elevation. The wind seems to howl constantly, sending its icy fingers probing around window frames, under door jambs, and straight into your bones. A single thought seems to travel up and down the canyon during these times, to each poor soul huddled in private misery while listening to trees crack and topple in the gale: "This wind is driving me #@^%ing nuts! There has *got* to be a way to use it."

Back in the day, there were very few commercial small wind turbines even available, and all were financially out of reach for everyone in our tiny, remote off-grid community. Strange contraptions rose up from junk piles behind cabins and spun fitfully in the wind. Many builders used surplus computer tape drive motors with blades stuck on the shaft. Others tried car alternators with belts and pulleys connecting them to the blades, and in the next canyon over was a huge gadget built with halved barrels swinging precariously like a giant anemometer. One monstrosity was even erected employing cloth wings that opened and closed as it spun—it looked exactly like a giant pterodactyl that had made an unfortunate crash landing and skewered itself on a fence post, flapping in its death throes. After days, weeks and months of labor in building all these machines, only the best could generate any useful amount of energy at all. Then they started failing one by one, as Nature had her way. Most lasted under a year before disintegrating.

In short—nobody up here had even the slightest idea of what they were doing. We had no clue about how wind power worked or how to build a successful turbine to harvest it. *The Mother Earth News* was our biggest source of information, showing giant homebrew turbines with cloth-covered sails and old waterpumper windmills fitted with car alternators. *Home Power Magazine* was still a fledgling publication. "Google" still meant 10^{100} and there seemed to be no way for the newly-hatched "information superhighway" to ever make its way up our narrow and winding dirt road. There was talk of someone in New Zealand named Al Forbes using exotic magnets to build successful and efficient alternators that could be used for wind power. And there was buzz about some guy in Scotland named Hugh Piggott who was building excellent wind turbines out of junk from the local scrapyard. But only a trickle of that information made its way across the big pond to our remote canyon.

Flash forward a decade. Inexpensive satellite internet became available, and the trickle of information turned into a flood. Co-author Dan Bartmann started an online junk store, and big surplus rare-earth magnets became his top seller. Everyone in the canyon had big junk piles full of the raw materials of mechanical and electrical engineering—scraps of steel, tangles of copper wire, and numerous vehicles that no longer ran. Dan started building alternators and wind turbines from scratch and hired writer and photographer Dan Fink shortly thereafter to help start up a new website, *www.Otherpower.com,* to promote the idea of do-it-yourself renewable energy. "Make your electricity from scratch" and "The cutting edge of low technology" were (and still are) our mottos. Instead of charging for the information, we put it up for free and simply provided a convenient place to buy the needed parts to build our projects.

The initial web pages we put up about homebrew wind power are still there on our website for all to see. At first the turbines were (to put it kindly) crude, inefficient, and prone to spectacular failure. But shortly thereafter, we met Hugh Piggott at one of his homebrew wind power seminars, and started absorbing his knowledge. We kept going back, for each and every one of his subsequent US seminars. Hugh was kind and patient with us, preferring to share his knowledge rather than to make a quick buck (er, well, Euro) with it. Over beers late at night, he was willing to inculcate us with the black arts of designing a furling tail, with how to make alternators that were (reasonably) efficient and (usually) didn't burn out, and with how to build a strong frame where bits didn't (frequently) fall off.

We started building wind turbines—dozens of them. We convinced numerous neighbors that "You really do need a wind turbine!" and "Why not come up to our shop and build it?" The mechanical parts were scavenged from the many old Volvos that seem to proliferate in our canyon. And thanks to the addition of Hugh's knowledge, these turbines worked pretty darned well. Many of them are still flying today! Then *Back Home Magazine* published our four-part series of articles on how to build a wind turbine from salvaged Volvo parts, and our websites kept growing with folks from all over the world joining up and building wind turbines.

This steady production of turbines for neighbors soon depleted our junkyard stash of car parts, and we began having the various critical bits made from steel with a CNC water jet cutter. That was a big improvement, at least as far as the aesthetics of these machines went—they weighed less and looked a bit more professional. We kept making refinements in the design, and started building both smaller and larger turbines—an entirely new proposition altogether, since wind turbines don't yield willingly to scaling up or down. To date we have built over 100 of the 10-foot diameter turbines described in detail in this book, and we feel very confident that if installed and maintained properly, they will rival any commercial small wind turbine of similar size when it comes to efficiency, reliability and noise—and the dedicated reader can build one for a fraction of the cost of a commercial product. The new 7-foot and 17-foot designs that we briefly cover here are coming along nicely and holding up to the abuse of Nature, with more and more going up every month.

But, now the bad news for the reader: *This book is already obsolete!* Yes, that was not a grotesquely large typo. And it's the main reason that this book took almost three years to go to press. We'd have a chapter ready to be put to bed, and Dan B or Rich or George in the windmill shop would suddenly say "Hey—I found a great new way to build this part. It's faster, it costs less, and the result is stronger and looks better." After thanking them profusely, Dan F would morosely wander home, pull up that entire chapter, and hit "delete." We solved some of those issues by adding sidebars to the book, others needed a complete re-write.

So, consider this book a work in progress, and remember that it will always be so. We'll be posting updates and corrections to the book, notes and photos from readers, and miscellaneous rants and raves about wind turbine design and construction on a special website created just for the purpose, *www.homebrewwind.com.* And if *you* come up with a better way of doing something, be sure to let us know. It might make it into our next book.

Have fun with the wind! DAN BARTMANN and DAN FINK

1. Introduction to wind power

We admit it—we are both wind nuts. We'd rather sit in lawn chairs outside in a blizzard, watching our turbines flying in the gale, than sit inside and watch television. And since we both live 12 miles from the nearest electric pole or phone line, the extra energy input is much appreciated. Here in the Northern Colorado mountains, if the sun isn't shining it's usually windy, so wind power makes an excellent complement to solar power. Building wind turbines is just as entertaining as watching them—simple enough to build in the home workshop with minimal tools and equipment, but mysterious enough that one could spend a lifetime experimenting with new ideas and designs and still not achieve the "perfect" wind turbine.

Do you need a wind turbine?

If you live off the grid in a rural area with a renewable energy (RE) system like we do and have a location that's friendly to wind power, a 10-foot diameter wind turbine like the one in this book could make a tremendous difference in your energy situation and reduce the runtime of your backup fossil fuel generator significantly. Every watt-hour you gain will be on the positive side of your energy ledger.

If you live in the city or suburbs, you might not even be *allowed* to raise a turbine high enough to make significant energy. And if you want to connect it to the utility grid to reduce your electric bill, you might be surprised at how much electricity you use, how much you waste, and how little energy any "small" 10-foot turbine can make in comparison. Plus, it's expensive both to buy the proper equipment to tie into the grid and to obtain the proper permits and inspections required to do so.

The Americans in a typical home consume 750 kilowatt-hours (kwh) of electricity per month. Kilowatt-hours is the figure you see on your electric bill next to the scary dollar amount you owe. But grid electricity is cheap and heavily subsidized compared to renewable energy—at least right now, in 2009. Unless you have local, state or federal incentives available and a favorable buy-back rate from your local utility, it could take decades to make your renewable energy investment money back. And a good part of that investment would most likely be going towards making electricity that you immediately waste!

NOTABLE QUOTES
"Houses don't use electricity, people do!"
Victor Creazzi, AeroFire Wind Power

If you live on the grid, you should do your homework and put money into conservation before you even start to look at investing in a renewable energy (RE) system.

It's estimated that for every dollar you spend on energy conservation, you'll save $3–5 on the cost of the RE system needed to power the stuff in your home. Inefficient appliances and lighting, phantom loads and bad habits can waste more energy per month than many RE systems can even contribute per month. For example, buying a brand new, efficient refrigerator will usually cost you much less than the cost of the solar panels, wind turbine and extra battery capacity to run your 20-year-old model. Electric heating elements like those found in space heaters, water heaters and ranges use so much electricity that they are impractical for most RE systems—it would be far cheaper to convert to direct solar space heating, a gas range and a direct solar or gas water heater than to try and power that equipment with solar or wind-generated electricity.

People who live with grid power usually don't even realize how much electricity they waste every day. You'll rarely see an off-grid dweller leave two televisions and roomfuls of lights on if nobody is using them. You can easily track your on-grid home's energy usage by simply reading your utility bill each month. It will show how many kwh you used and what they cost you. You can also buy smaller kwh meters that track the energy usage of individual appliances and circuits. Commercial wind turbine manufacturers provide estimates of kwh per year output from their turbines at different average wind speeds so you can get an idea of what to expect coming back in. The comparison of your energy consumed versus what a wind turbine can produce may be very sobering!

DOG HAIKU
Television on
No humans watch, only dogs
Power is wasted

Do you have a good site?

Wind turbines need to fly high in the air—at least 30 feet above any obstacle within 300 feet—to live up to their potential. Trees, buildings and terrain slow down the wind significantly, and installing a wind turbine where the wind is slow is like installing a solar panel in the shade! Such obstacles also impart turbulence to the air, and turbulence both hurts turbine output and increases stress on the machine.

Generally, a good tower for your wind turbine will cost *at least* as much as the turbine itself. Getting 30 feet above any obstacles within 300 feet usually means a very tall tower, so you'll have to check local regulations and building codes, talk to your neighbors and possibly get permits. As an example, our local county building codes require a permit if the tower exceeds 40 feet, and the "fall zone" of the tower must lie entirely

> **NOTABLE QUOTES**
> "Although people don't like to hear the message, the truth is that the cheapest way to increase the power output of a wind generator is to increase tower height."
> *Mick Sagrillo, Sagrillo Light and Power*

on your property. In some locations 40 feet might be ample for flying the turbine above obstacles, but in a wooded area it will likely not be high enough. Wind power is best suited for a property of at least two acres in a rural or semi-rural area.

The authors often "fudge" the 30/300 foot rule. In the steep, rugged mountains where we live, good tower sites are scarce and may not even be on the right piece of property. Tilt-up towers are tricky on uneven terrain, and we don't like to climb fixed towers. But remember—we live completely off the grid. So we appreciate any possible energy gain, and we are not trying for any payback of utility bills. In a grid-tied situation where the reduction of utility bills is the goal, it's a silly waste of money to fly a wind turbine on a tower that's too short. Off the grid you can be a bit more flexible.

Don't even *think* about mounting a wind turbine on your roof! Buildings are *not* designed for the extra load a wind turbine can impart—it could damage your house. Even worse, wind speeds near roof level are very low, and perform-

> **NOTABLE QUOTES**
> "Rooftop wind turbines are a load of nonsense."
> *Hugh Piggott, Scoraig Wind Electric*
>
> "Mounting wind turbines—of any kind—on a building is a very bad idea."
> *Paul Gipe, Wind-Works*

ance will be disappointing. It's far better to try and follow the 30/300 rule to maximize the "fuel" your wind turbine sees. Winds near rooftop level are also very turbulent, hurting both energy output and reliability.

Measuring the wind

The more time and money you plan to spend on wind turbines, the more time and money you should spend figuring out your wind resource potential. When most people try to envision what the average wind speed is at their site, they mentally consider only the times when the wind is blowing. The official figure of "average wind speed," which turbine manufacturer's estimated kilowatt-hour per year output figures are based on, also takes into account all the hours during which the wind is not blowing at all. This is difficult to estimate, and such estimations usually end up overstating the wind potential significantly. This is a problem, because the difference in a turbine's output at (for example) 10 mph and 12 mph wind speeds is very large. A typical 10-foot diameter turbine can make 75-100 kwh per month with a 10 mph average wind speed, and 125-150 kwh per month at a 12 mph average wind. That's a big difference!

One way to estimate your average wind speed is to fly a logging anemometer (Figure 1.1) at your site, at planned tower height, for a year. This is not possible for many folks, including us—a good logging anemometer costs around $400, and a tall tower to fly it on will cost even more than that. And what if it was a particularly calm or windy year? If you are thinking about making a significant investment of money in wind power, you should also check government wind speed data from the National Renewable Energy Laboratory (NREL, your tax dollars at work) to get a sense of whether or not your location is windy enough for a turbine to be feasible. Rough maps from NREL are included in *Appendix F*, and there are more detailed free sample map downloads available at *www.AWStruewind.com*.

Figure 1.1—APRS World cup anemometer and data logger for measuring wind speed.
Photo courtesy of APRS World LLC

Simply talking with neighbors and observing the terrain and vegetation around your potential tower site can be very valuable. Ridge tops are usually excellent, and trees that show branches that have been stunted by prevailing winds (called "tree flagging" and explained in *Appendix F*) can be a rough indicator of average wind speed. Another option is to build a homemade logging anemometer, or buy an inexpensive version that does the logging via PC software instead of stand-alone hardware.

Average wind

The statistics used to calculate the distribution of wind speeds are complicated, but the results are easy to understand. Most wind comes to us at low or moderate speeds, and higher winds are relatively rare. In most locations worldwide, the distribution of wind speeds keeps fairly close to a Rayleigh distribution, shown in Figure 1.2. There are non-Rayleigh locations where the curve takes on other shapes, but these are relatively rare. The distribution shown here is relatively common.

In Figure 1.2, the horizontal axis of the graph shows wind speed, and the vertical axis shows the probability (which can be condensed down to the predicted number of hours per year) that the wind will be blowing at that speed. The area under the curve is always equal to the number of hours in one year, because the probability that the wind is either blowing or not blowing is 100 percent. Some logging anemometers can even plot your location's wind speed distribution for you, and PC software can do it too.

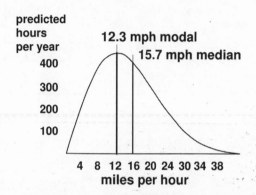

Figure 1.2—Probability distribution of how the wind usually comes to us. The "modal" wind speed of 12.3 mph is simply the most common wind speed during the course of a year. At the "median" wind speed of 15.7 mph, half of the time the wind is blowing slower than the median and half the time it's blowing faster.

Rated output

Rated output is overrated! But almost everyone interested in installing a wind turbine asks about the rated power output of different brands, many uninformed people comparison shop for wind turbines by rated output, and unscrupulous advertisers even try to compare wind turbines to solar panels using rated output. *It's all hogwash.* Rated output is an almost worthless figure—it's simply the maximum sustained output that the turbine can produce. But at what wind speed does it make that rated power? The rated outputs of reputable manufacturers are taken at 28 mph or below. And for how many hours per year do you get wind at such a high speed? In most locations, very few—take another close look at Figure 1.2. Most winds come in at a much lower speed, and therefore have much less power available in them. There are only two possible

CRUCIAL DEFINITIONS
Swept Area
The area of a cross-section of moving air that a wind turbine rotor can extract power from, in square feet or square meters. The formula is **pi × r²**, where pi=3.14 and r=the rotor radius (the length of one blade). For a 10-foot diameter rotor, the swept area would be $3.14 \times 5^2 = 78.5$ square feet.

Rated Output
The maximum *sustained* output a wind turbine can produce in high winds. Reputable manufacturers rate their turbines at 28 mph or less.

ways to compensate for this: a bigger swept area to capture more wind, or a higher tower to increase average wind speed. Shop for wind turbines by swept area, *not* rated output.

Folks often ask us what the rated output of the wind turbine design in this book is. We like to call it 700 watts, or just tell them, "It's a 10-foot turbine." A reputable commercial wind turbine manufacturer might rate this turbine at 1,000 watts (1 kw). A less reputable one might call it 1.5 kw, since it could possibly reach peaks of that output during a severe storm. A really disreputable manufacturer (there are a few out there, buyer beware) would rate it at 60 mph and call it a 25 kw machine! By saying 700 watts we are hoping that folks who build and site this turbine correctly will occasionally be impressed that it can actually hit a kilowatt of output in very windy conditions, while understanding that rated outputs are useless in the first place. Saying, "It has a 10-foot diameter rotor" is much more meaningful information.

Shopping for a wind turbine design by swept area also makes sense when you consider how you normally pay for electricity from the utility—by kilowatt-hour, not by kilowatt. The larger your swept area, the more kwh you can make per month. Wind turbine manufacturers can also provide you with a rough estimate from their machines in kwh per month, and you can connect a meter to your new turbine to measure it yourself also. If you compare these figures from different turbine manufacturers, you'll find very little difference in kwh per month output between machines with the same swept area, while rated output might vary considerably! Rated output has almost nothing to do with the energy generation potential at your site.

Many folks also ask us why we don't just build a whole bunch of small, lightweight wind turbines instead of these 10-footers. Flying four small 5-foot turbines will not give you nearly as much energy output per month as flying a single 10-foot turbine. As you'll learn in the upcoming chapters on wind power theory, in low winds those tiny turbines will be sitting idle while a larger one will already be making power. And, since a tower usually costs at least as much as the turbine flying on it, it's much more cost effective to fly a single, larger turbine than multiple smaller ones.

Noise, neighbors, birds and bats

As soon as you start telling your neighbors about your plans for a wind turbine, at least one of them will likely be worried and ask about the noise, bird kills and bat kills they think will surely result from your project, and be greatly concerned about you ruining their view.

There are some fairly noisy small wind turbines on the market, but the design in this book is not one of them. Small, inexpensive commercial turbines are often lightweight and fast-spinning, and can be somewhat noisy—that's just one of the trade-offs manufacturers use to keep costs down and prices low. Ironically, some of the noisiest wind turbines on the market have model names that imply how quiet they are! The designs in this book instead follow the "heavy metal" philosophy of turbine building, as do many of the more expensive, quiet and reliable commercial brands. These turbines are heavy and spin slowly, so therefore they don't make as much noise.

There is always some noise emitted from any wind turbine—the gentle whooshing of the blades and a low hum when the alternator is making power. Most of the time, the sound of the wind blowing through the trees is far louder than any noises the wind turbine itself makes. We often describe the noise to folks as being about the same as someone riding by on a bicycle.

If your neighbors will be able to see the completed turbine on its tower, calmly discuss the project with them. The authors think wind turbines are beautiful things to watch in operation, but others may disagree. Simply picking a color scheme that blends into the skyline can make a big difference in aesthetics. Another issue for your neighbors might include a few minutes of "strobing" in the morning or evening for a couple weeks during spring and fall, when then sun moves behind your wind turbine blades from the neighbor's perspective.

Bird kills have been heavily hyped in the media for decades and bat kills were discovered relatively recently. All of these cases involved giant commercial wind farms, not small turbines. In short, some of the first big multi-turbine wind farms

> **NOTABLE QUOTES**
> "There is no evidence that birds are routintely being battered out of the air by rotating wind turbine blades as postulated by some in the popular press."
> *Mick Sagrillo, Sagrillo Light and Power*

were poorly sited along migration routes. Headlines called wind turbines "Raptor-Matics" and "Cuisinarts of the Skies." In reality, the biggest killers of flying animals by far are buildings, domestic cats, pesticides, habitat destruction, power lines and communication towers.

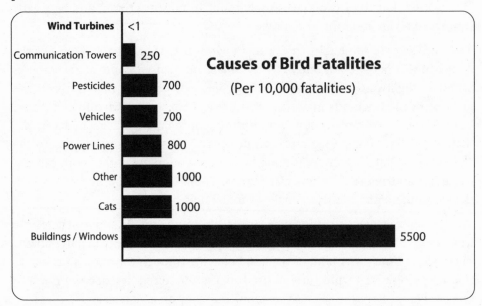

Figure 1.3—Causes of bird mortality.
Source: Wallace Erickson, et al, 2002: Summary of Anthropogenic Causes of Bird Mortality

The fact is, *any* tall, man-made structure kills birds! Wind turbines have simply received a bad rap in the media, while bird kill problems from tall buildings and ubiquitous cell phone towers are ignored. Figure 1.3 shows the real story.

Extensive studies of wind turbine animal kills have revealed that, when proper studies of migration routes are undertaken before choosing a site, the utility-scale turbines themselves cause no more flying animal mortality than any other tall, man-made structure. We have erected dozens of small wind turbines over the last few years, and have yet to record a single bird or bat kill. It's most likely because of the size—the small-scale wind turbines we build are dwarfed by even a typical cell phone tower.

CAT HAIKU
Nap under turbine
Blades spin, but no birds are killed
I was so hungry...

2. Renewable energy 101

If you are seriously considering building a wind turbine, you probably have already installed a functional renewable energy (RE) system. However, you might only be in the planning stages for your system or new to renewable energy entirely—if so, this brief chapter is for you! Whole books have been written about what we're going to cover in only one chapter. We highly recommend that you check our *Sources* chapter and get some books for further reading, plus a subscription to *Home Power Magazine*.

$$\text{Power} = \int \frac{\sum \psi \neq x^2 \xi^{42}}{z^3 \omega \Rightarrow \phi\, \psi}$$

Figure 2.1—Simple renewable energy math. Memorize this formula, and…We're just kidding! RE math is quite painless, and we'll present it in small bites.

Almost anyone could easily and very successfully design, install and use an entire home electrical system without understanding a whit about electricity. As long as you know what it *does*, who cares what it *is*, right? Absolutely! However, a basic understanding of some of the concepts and terms involved will (hopefully) clear up some of the more mysterious things that happen inside an RE power system—and help explain why successful turbines and systems are designed the way they are. Our sidebar that attempts to explain basic electrical theory begins below. Skip it if you wish, but it might help clear some concepts up for you.

What is electricity?

There is no good definition for the word "electricity" because the same word is used to describe so many different things. Instead, it's much more accurate to talk about the three concepts and terms that we'll be working with most in renewable energy, "electrical charges," "electrical energy," and "electrical power." These are *all* commonly called "electricity" by most folks, and by not making this distinction great confusion results—one reason why electricity is a dreaded subject for many students.

Electrical charges

Electricity is a basic component of all matter—oops, darn it—better to say that electrical *charges* are a basic component of all matter! They are called protons and electrons, and every atom of everything has them. Protons are defined as having a positive charge, electrons have a negative charge—they are opposites, and they attract each other. Chunks of charges are measured in coulombs. In most substances, these charges mostly stay put. Metals, however, are very unique as far as their electrical charges go—the charges are easily movable. Inside a piece of metal or in a wire the electrons are floating around randomly in a big "sea," while the protons stay put with their atoms. Metals are called "conductive" when they have these properties. Electrical charges are actually visible—this sea of charge is why metals look shiny.

Now, note from the explanation above that your metal (let's say copper wire) is already stuffed full of charges! You can't make any more, and you can't destroy the ones that are already in there—that's the first law of physics. What you *can* do is force the charges to move. The charges also need somewhere to go, however, and the only option is back to where you started pushing on them—that return path is what's called a "circuit," and the charges in your wire won't move very far without it. That seems to make sense—a flashlight battery that's not hooked up to a light bulb will just sit there, same with an electric heater or television that's not plugged into the wall.

Amperes

That brings us to the first important unit for electrical measurement, amperes (amps, shown as "I" in electrical math). Sometimes amps are referred to as "current," like the current in a river. Remember that we mentioned earlier that quantities of electrical charge are measured in coulombs? Well, one ampere equals one coulomb of electrical charges moving past you in a circuit for one second. After they slowly move past you, they continue right back to their source, via the other side of the circuit. If you pretend that the charges inside your wires are water, then coulombs of electric charge are analogous to gallons of water, and amps are analogous to gallons per minute of flow. You can't store amps, just like you can't store gallons per minute—they are both a rate of flow, *not* a substance. You don't dip out a bucket of current from a river, you dip out a bucket of water!

Charge pumps

There are many ways to push these charges around, though here we'll cover only the three most common examples that occur in an RE system. One possibility is that you can move a magnet past a metal wire—that's what happens in a wind turbine, a hydroelectric turbine, a gasoline backup generator, or a utility power plant. Or, you can touch two different varieties of conductors together and shine a light on them—that's how a solar photovoltaic (PV) panel pushes those charges around. And, if you separate the positive and negative charges from each other, they immediately want to move back together because the opposite charges attract each other. So, if you forcibly hold those charges apart, you have a storage battery. Figure 15.2 shows a simple circuit. The charge pump is your battery, PV panel or wind turbine, and the load could be a light bulb or electric heater.

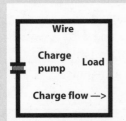

Figure 15.2—Amperes are the rate of flow of charges in a circuit, and are equal to coulombs per second.

In those three examples above, notice how in each one you have to perform some work to push the sea of charges around? You have to physically move the magnet past the wire, the light particles have to slam into the solar cell, or you have to physically separate the charges in a battery.

Electrical energy

And that's the next key concept—the difference between electrical charge and electrical energy. The electrical charges are always there in your wires, and stay put until you move them. The electrical energy is what *you* put into making the charges move, and is measured in chunks called joules. You can use joules to measure the energy made or used by anything—your legs spinning a bicycle wheel, your arms lifting a weight over your head, the wind pushing on your wind turbine blades, or the energy used by an electric heater. Your wind turbine, battery and PV panels are really just "electrical charge pumps." Your electrical utility sells you electrical charge pumping services, not electricity!

There are a few funny things about this electrical energy you just made by pushing on the sea of electrical charges: for one, it moves fast, at just under the speed of light. Secondly, it doesn't move inside your wires, but instead in the space *around* the wires! Lastly, this energy you put in gets used up (unlike electrical charges which always remain in the wire)—your electric heater or light bulb sucks the electrical energy right out from around your house wires and converts it into light and heat, without sending any electrical energy back to where you pumped it from.

How can this energy travel so fast? Well, try a spinning bicycle wheel for an analogy—pretend that the rubber in the tire is your sea of charges. Spin the wheel with your charge pump (your hand in this case), and the charges move. When you do this, the other side of the wheel starts moving at exactly the same time that your side started moving—nothing mysterious about that. But the rubber of the tire (the sea of electrical charges) itself is moving only slowly past you! It's the same with your charge pump—all the charges in the circuit start moving all at once, but are moving slowly past you. The energy *you* put into the circuit moves instantly.

How does this energy get used up while the charges do not? Well, press your finger against the spinning tire. Your finger gets *hot* from the friction, and the wheel quickly slows to a stop. You just transferred the energy you put into spinning the wheel right to your finger, which used it up by converting it to heat. It doesn't send any energy back, though the rubber in the wheel is still moving right back to the charge pump. Now spin the wheel with one hand, while holding your finger on the other side of the wheel with the other hand. You have to push hard to make it spin, and the harder you push the hotter your finger gets. Push harder with your finger, and the harder you have to pump the wheel, and again the hotter your finger gets. In an electrical circuit, your finger is analagous to a light bulb—and things that take electrical energy out of the circuit and use it up are called loads.

Watts, ohms and electrical power

We've just described watts, watt-hours and ohms with this last example! Ohms are the resistance in the circuit—how hard you are pushing your finger down on the slowly spinning rubber (the charges) in the wheel. And watts are joules per second—how hard and fast you are pumping that wheel to transfer energy to your finger. Push on that wheel for a couple minutes, and you've put a certain number of watt-hours into the circuit, and taken it right out again with your finger converting it to heat. Watts are a measure of electrical power, and watt-hours are a measure of electrical energy.

Volts

The last thing remaining to define is voltage, and then we can discuss why power moves energy instantly, and also moves it outside the wires, not in them. Let's say you put some electrical energy into your circuit via the charge pump (perhaps a PV panel this time), and it starts making the negative electric charges in your wire move, while the positive charges stay put (the characteristics of a conductor). Now that you started the pump, one side of your circuit has a whole bunch of negative charges, and the other is left with mostly just positive charges—there's a charge imbalance. This difference between the two sides of your circuit (more negative on one side of your charge pump, more positive on the other) instantly formed an electric field around the wires when you turned on the charge pump, and the energy you put into that field is what made the charges want to move. That field *is* voltage. It's called the E-field and if it was visible, it would look like Figure 15.3.

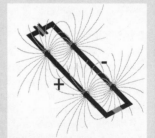

Figure 15.3—The electric field (E-field) around a circuit's wires is measured in volts.

A "field" means that there's a gradient involved, stronger near the source and weaker as you move farther away. The more electrical energy your solar panel charge pump puts into the circuit, the bigger the charge imbalance and the harder the field pushes. And the harder it pushes, the greater the imbalance of charges (voltage) becomes and the faster the charges want to move back to where they started, at the other side of the pump. If you could see the electric field around a flashlight battery sitting on your table, not connected to anything, you'd see an electric field around the battery, connecting the + and – ends in space!

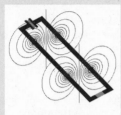

Figure 15.4—The magnetic field surrounding wires in a circuit when power is applied is called the B-field.

There's another field around your wires besides the electric one created by voltage. It's a magnetic field called the B-field, and it's caused by the motion of the charges in your circuit (amps) instead of the difference in the charges (volts). If you could see this magnetic field, it would look like Figure 15.4—and if you sprinkled iron filings around your wires while energy was moving, they would align themselves with the field, allowing you to visualize it.

It's the combination of these two fields that lets power (watts) transfer energy (watt-hours) in the space around your wires, and it's commonly called an electromagnetic field (EMF). This is all exactly the same phenomenon as happens with radio waves or light rays, but it happens in the space all around the copper wires. This electromagnetic field is "pulled in" towards the wires and then "sucked" into the light bulb filament as it transfers energy. What pulls it in? The sea of charges in the metal wire—the more charges that are pulling at the field (in other words, the thicker the wire), the more energy it can move at one time (watts), and thus the more energy it can can move *over* time (watt-hours). If there aren't enough charges to pull in and direct the electromagnetic field to the load (in other words, if the resistance in ohms is high), more of the energy that you put into the circuit is wasted as heat.

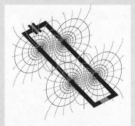

Figure 15.5—The combination of the E-field (volts) and the B-field (amps) around wires in a circuit allows power (watts) to transfer energy (watt-hours) to a load.

Figure 15.5 attempts to show what this combination electromagnetic field would look like around a circuit while it's transferring energy to a load, if the EMF was visible. For clarity, only two pairs of field lines are shown—if you could really see both the E-field and the B-field, each wire would be surrounded by an entire "tube" of fields, like a big ring of sausage.

Ohm's law

You'll often hear about Ohm's law when working with RE systems. In 1827 Georg Ohm published a paper about the relationship between electrical pressure, flow, and resistance, which was immediately dismissed as nonsense by his peers in the scientific community. It was not widely accepted until 20 years later—and today, it's the most important math used in any electrical system. Almost all of the math involved in designing and building both RE systems and wind turbines comes straight from Ohm's law. You are likely to use it frequently to calculate how thick the wires in your power system must be so they don't overheat, how much energy your battery bank can store, and for how to design a successful wind turbine. And for everday reasons too, for example to find out "For how many hours can I watch *Dukes of Hazzard* re-runs on my 100-watt television with this size battery bank?" In this text we group Joule's power law (which gives us watts) with Mr. Ohm's law for convenience and because they are so often used in combination. Joule's law is easily derived from Ohm's law.

With the examples above, we've actually already discovered Ohm's law. Really! All of the variables needed have been covered—volts, ohms, amps and watts. Let us explain…First, notice the symbols we'll use in the math: **volts=E, ohms=R, amps=I, and watts=P.** And the basic equation of Ohm's law is that:

voltage / resistance = amperage, or **(volts / ohms = amps)**, or **(E / R = I)**

You can translate this into English as: *The harder you push, the faster the electrical charges move. Make the voltage twice as high, and you get twice the amperage (twice as many coulombs per second move by). And inversely, it means the more resistance, the slower the charges move.*

Now flip the equation around:

amps × ohms = volts, or **(I × R = E)**

You can translate this as: *A flow of charges produces a voltage if it encounters resistance. And (if resistance stays the same) the more current, the higher the voltage you get. Or, if the current is forced to stay the same and you increase the resistance, then the higher the voltage.*

What is power?

Power (watts) is important in RE systems because it measures the rate at which energy is moved to and fro, and you can then just multiply by time to get watt-hours to measure energy. Power will measure both the rate at which electromagnetic energy is produced (output from your wind turbine and solar panels) and the rate at which your loads consume it (light bulbs, electric heaters, and psychedelic lava lamps). So we start with:

volts × coulombs = joules

This means that: *it takes energy to push charges against voltage pressure. But charges are not really what we care about, it's the* **flow** *of the charges that transfers energy! So:*

volts × coulombs per second = joules per second

You can translate this as: *it takes a flow of energy to make charges flow against pressure.*

Hey, wait! Coulombs per second are known as amps, and joules per second are known as watts. So we get:

volts × amps = watts, or **(E × I = P)**

This can be translated as: *pushing charges against a voltage requires power.* It also explains why you can't charge a 24 volt battery with a 12 volt solar panel—you need to push with at least 24 volts to make anything move, and you have to put power over time (energy) into the circuit to do that.

That was the beginning of Joule's law, here's the rest, after combining it with Ohm's law:

amps² × ohms = watts, or **(I² × R = P)**

Where did the squared come from? Remember the formulas for volts:

(volts = amps × ohms),

and amps: **(amps = volts / ohms)**,

and then watts:

(watts = amps × volts).

So you could also write the formula for watts as:

(amps × ohms) × amps = watts, and also: **(volts / ohms) × volts = watts**

These would translate as: *voltage applied across ohms uses up a constant flow of electrical energy (power, measured in watts),* and: *when charge is flowing against resistance, a constant flow of electrical energy (power, measured in watts) is being used up.*

The really important parts of these last examples to understand are that: when you double the voltage *or* the amperage, energy flow (power, in watts) increases by a factor of four (the squared part of the equation). Changing the resistance doubles or halves the power.

I² R losses

And all this math above also explains why higher-voltage electric systems and wind turbines are preferable—you want to get as much energy as possible from your solar panels or wind turbine into your RE system, and as much energy as possible from your system to your loads. Any losses from resistance are turned into heat in your wires, and are not available for your use! Because of that "squared" in the equation, you can send the same rate of energy flow (power in watts) four times as far on the same piece of wire (ohms stay the same) if you double the voltage. Or, inversely, you'll have to use wire that's twice as thick (and very expensive) to keep the losses (ohms) the same if you are running at half the voltage. This becomes a big factor in wind turbine design, too, as you'll learn later in Chapter 20, *Scaling it up and down.*

We hope this diversion into electrical theory has explained some of the mysteries involved with renewable energy systems. If not, don't worry! Just do the math when designing your system, and everything will work out fine.

There are some easy tricks you can use to avoid doing algebra every time you try to use Ohm's law. The best is called the "Ohm's law wheel," and breaks it all down into every possible combination of equations. It's shown at right in Figure 15.6.

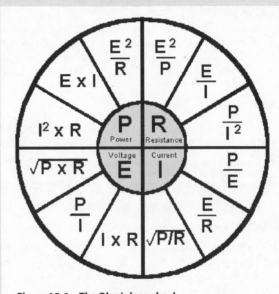

Figure 15.6—The Ohm's law wheel.
Drawing courtesy of Lewis Electric Co.

Many thanks to Bill Beaty for the diagrams and help with this section about electricity. Bill is a research engineer at the University of Washington, and devotes much of his time to explaining electricity so that even his grandmother can understand it. He also runs the coolest science website ever in his spare time (www.amasci.com) and he really enjoys exposing errors in school science textbooks.

DOG HAIKU

Volts, ohms, amps and watts
Dog power is measured in
kibbles per hour

What comprises a renewable energy system?

All RE systems have three elements in common:

Energy sources: Devices that generate electricity, such as wind turbines, photo-voltaic panels, micro-hydro plants and fossil fuel generators.

Energy storage: A way to store energy for use when there is none coming in.

Loads: All the lights, appliances, and gadgets that you run in your house.

RE systems can be classified into three types:

Off-grid: The system is not connected to the grid (electric lines from the utility company). Energy is stored in a bank of batteries that run the loads. Excess energy is used for air or water heating.

Grid-tied: The system is directly connected to the grid. Energy is not really stored, instead the grid is used as a "battery." If the grid experiences a blackout, there is no energy for loads. Excess energy is sold back to the utility.

Grid-tied plus battery backup: The system is connected to the grid. Energy is stored in a battery bank, so if the grid experiences a blackout the house still has energy to run loads. Excess energy is sold back to the utility.

Grid-tied homebrew turbines?

Wind turbine homebrewers almost always build their turbines for off-grid use only, but some enthusiasts have successfully connected their creations to the grid. One big issue with grid-tying homebrewed turbines is the expensive extra equipment, utility company and government regulations, inspections and permits that are required to connect to the grid. You may have trouble getting a permit to install a homebrewed wind turbine, and it won't be UL listed—something your building inspector may require. However, at the time of this writing there *are* no actual UL standards for commercial (or homebrew) small wind turbines, so they cannot be UL listed in the first place—only the subsequent electrical system to which they are connected can be. This is an essential safety issue—during a grid blackout, power company workers will be handling the power lines, making repairs. They could be killed if your improperly designed or installed grid-tie system made the lines live accidentally.

It will be *far* easier for you, both technically and politically, to connect your homebrew turbine to a battery-based off-grid or grid-tied power system rather than a direct grid-tie system.

Everyone in our little community up here in the mountains lives entirely off grid, 12 miles from the nearest power line. We don't have any experience with designing, building or installing grid-tied systems. However, other homebrewers around the USA have successfully tied this wind turbine design to the grid. If this is your intention, be sure to discuss it with your local building code inspector, your local utility and hopefully a local RE company that has installed grid-tied systems in your area. The turbine itself would wire into your grid-tied system just like any other commercial wind turbine of similar size. Also, be sure to read some books on grid-tied applications—there are some listed in our *Sources* chapter.

The wind turbine design in this book is intended **only** *to charge the battery bank of a renewable energy system. It will* **not** *function properly if connected to run heating elements directly, nor for a direct grid-tie application with no battery bank. At this time, we and other homebrew wind power enthusiasts are experimenting with new alternator designs for direct grid-tie applications, but we do not have any data yet.*

It is your responsibility to install any power system, off or on-grid, to comply with all code regulations, and to obtain all required permits and inspections. An improperly designed RE system could also cause a fire. If you don't know what you are doing, hire an electrician!

Choosing the system voltage

When designing a renewable energy system, choosing the system voltage is one of the first decisions to make. In the early days of off-grid RE, 12 volts was the most common because many systems drove 12 volt DC (VDC) loads directly, and inverters to convert 12 VDC power to 120 volt AC (VAC) "house current" were large, inefficient and expensive. 12 VDC appliances and efficient lights were (and are still) expensive, but less so than a large inverter at the time. As the price of inverters dropped, it became more common to have no 12 VDC loads in a house at all, and run everything at 120 volts AC from the inverter. That way common, inexpensive appliances and lights could be used. Systems with all-AC loads are the most common today, 12 volt systems are increasingly rare, and 48 volt and 24 volt systems are becoming the standard.

The reason is energy wasted as heat. The lower the voltage, the more energy is wasted, or the larger all wires must be to avoid this loss. Because of this, higher-voltage inverters tend to be more powerful for their size, too. Every other component and wire of a 12 volt power system must be larger (and therefore more expensive and harder to install) than its 24 or 48 volt equivalent, including the power semiconductors inside inverters and controllers. 24 and 48 volt inverters are now much more common and inexpensive, so it no longer has to be a huge investment to buy one.

Wind turbines also have big limitations at 12 volts. The internal wiring is more difficult to fabricate, thicker wires must be used in the long wire run from the turbine to the battery bank, and more energy is lost as heat in the rectifiers. Many commercial wind turbines are not even made in 12 volt versions anymore—they are found in 24 or 48 volt only. If you are starting from scratch, design your RE system for 48 volts! If you are already stuck with 12 volts and you can't afford to convert up, be sure to read Chapter 12, *Stator*, which includes some changes to this wind turbine design that we have used to minimize losses in 12 volt systems. You'll also discover that this 10-foot turbine design is about the largest you could practically install for a 12 volt system. Anything larger, and the line losses and heat become too great.

Sizing wires

By applying the combination of Ohm's and Joule's laws, you can now see why using a higher system voltage saves so much money! Wiring is rated by "ampacity," which is how many amps it can carry without overheating. Generally, RE systems are wired so that there is less than five percent loss from resistance heating in any leg of the system.

Take the output of a large 1 kw solar array and do the math:

$I = P / E$

amps = 1,000 watts / 12 volts = 83.3 amps

amps = 1,000 watts / 24 volts = 41.6 amps

amps = 1,000 watts / 48 volts = 20.8 amps

Chapter 18, *Raising*, gives a quick comparison of what wire sizes would be needed to run this wattage a certain distance. When designing a system, RE engineers don't run the calculator for every length of wire needed—the math is fairly complicated in figuring out what a five percent loss is. Instead, they use a computer spreadsheet or a wire sizing chart. You can also pull up wire sizing calculators on the internet that will do the job for you.

With wind turbines you can "cheat" on wire size a bit (Chapter 18, *Raising*) for the long line running from your turbine to the house, because a 1,000 watt turbine will only be making that maximum output on rare occasions—and because of some performance issues (also discussed in that chapter) that relate specifically to wind turbines and their alternators.

One more definition...watt-hours

In case you skipped over our explanation of electricity in the sidebar above, we stopped at defining power as being measured in watts. Watts are very handy, since everything can be measured using them—the power that a generator or solar panel produces at any given instant, the power that a television uses while you watch it, or even how much power you have to push with in lifting a rock over your head. The problem is, watts are still a measure of instantaneous power—how much power is being moved around at any given instant. There's actually another unit that we need to be concerned with, and that is "for how long was this power moving?" The units we use for measuring this are watt-hours, and they measure energy. A typical bill from the local electric utility will show how many kilowatt-hours (kwh) your home uses each month. One kilowatt (kw) = 1,000 watts, and one kwh means you've been using that 1 kw for one hour—or 100 watts for 10 hours. It really doesn't matter if your wind turbine or PV array was putting out 300 watts the last time you looked at it, or that your TV uses 300 watts. What matters is for how many hours you've been gaining or using that 300 watts!

Alternating current and direct current

Darn it, there's another set of terms we need to define before getting into the meat of an RE system. Direct current (DC) electricity is what we are familiar with from a car battery, and it "flows" in one direction only. Alternating current (AC) is what we are familiar with coming from the outlets in our homes, and its energy flow reverses back and forth in direction at a certain frequency, measured in cycles per second (Hertz, abbreviated Hz). In the USA, that frequency is 60 Hz. The important thing to remember here is that being exclusively DC devices, batteries can only be charged by DC current. The electricity from AC sources (like gasoline generators or most modern wind turbines) must be rectified to DC for battery charging. Photovoltaic (PV) panels, on the other hand, make DC directly.

Battery banks

Whole books have been written about selecting, sizing and installing battery banks for RE power systems. In fact, we recommend that you buy such a book and read it thoroughly! Check our *Sources* chapter in this book for some recommendations. The battery bank is the heart of any off-grid RE system, and also the most fragile component.

The batteries used for RE are not standard car batteries. RE ("deep cycle") storage batteries are necessarily large and heavy with thick, durable lead plates inside. All lead-acid batteries have individual cells of 2 volts each, grouped in combinations of 2 volts, 6 volts, or 12 volts per battery. Vehicle batteries are generally 12 volt, are lighter and have many more thin plates inside. They are made to give out large amounts of energy quickly for starting a vehicle engine, after which they are quickly recharged. They cannot be discharged deeply, and will be ruined if this happens more than a few times. Deep cycle batteries withstand this abuse better, though they will still be damaged and eventually ruined by too many deep cycles. A good rule is to never discharge the batteries by more than 50 percent of their capacity, though most will take an 80 percent discharge a few times without trouble. If you abuse your batteries in that way, be sure to charge them up afterwards both promptly and fully.

Battery capacity is generally measured in amp-hours, not watt-hours. This is a source of great confusion to some folks, and it would be much easier (and more accurate) for all involved to state it in watt-hours—but alas, it's not, so you'll have to use Ohm's (and Joule's) law to calculate it. All you need to do is multiply the amp-hour capacity of the battery bank by the system voltage to get watt-hours of capacity. This silly difference in terms is most likely a conspiracy to obfuscate the numbers so that RE writers and engineers can keep their jobs! "Cold cranking amps" (CCA) figures that are printed prominently on car batteries are not a measurement of battery capacity. What you are looking for in selecting batteries for an RE system are amp-hours of capacity, and you may have to refer to a battery spec sheet or ask your local battery dealer, as this figure may not be displayed prominently on the battery.

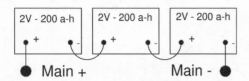

Figure 2.4—Batteries connected in series. 6 volts at mains, 200 amp-hours, 1,200 watt-hours.

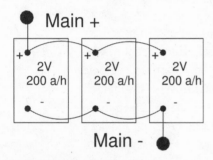

Figure 2.5—Batteries connected in parallel. 2 volts at mains, 600 amp-hours, 1,200 watt-hours.

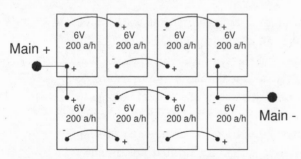

Figure 2.6—Typical home battery bank, connected in a combination of series and parallel. Two strings of four 6 volt batteries at 24 volts each, connected in parallel. 24 volts and 400 amp-hours at mains, 9,600 watt-hours.

Batteries in series, parallel, and combination

Since RE systems usually come in 12 volt, 24 volt and 48 volt varieties, how are 2 volt or 6 volt batteries incorporated? The answer is by wiring the batteries together in series, parallel or a combination of the two. It all goes back to Ohm's and Joule's laws, as does almost every bit of math used in designing and building renewable energy systems. Batteries wired in series give voltage that adds up when measured at the ends of the "string" of batteries, but the capacity in amp-hours does not increase, just the voltage. The capacity in watt-hours, however, *does* increase, because watts = volts × amps. In Figure 2.4 at left, three 2 volt batteries are connected in series. This is what would be found inside a typical 6 volt RE battery. Voltage at the mains increases, amp-hours of capacity stay the same, and watt-hours of capacity increase. Again: one watt of power to or from a battery for one hour = one watt-hour. 10 watts for one hour = 10 watt-hours. And 10 watts for 10 hours = 100 watt-hours.

Connecting batteries in parallel leaves the voltage at the mains the same, but increases both the amp-hours of capacity and (once again) the watt-hours of capacity. Figure 2.5 shows the same three batteries connected instead in parallel. Note that the watt-hour capacity remains the same as in the series connection above. However, a 2 volt battery bank is not very useful. How do you increase both the battery bank voltage and the capacity?

The answer is a combination of both series and parallel connections. A typical RE system battery bank is shown in Figure 2.6. Each series string of 6 volt batteries gives 24 volts at 200 amp-hours. The two strings are connected together in parallel to increase the amp-hour capacity to 400 a-h. The watt-hour capacity is 9,600 w-h, or 9.6 kwh.

Note that this series-parallel math holds true for PV panels also. For a 48 volt system that uses 12 volt PV panels, the panels are wired in series strings of four to make 48 volts, and each string is wired in parallel with the others to increase the amps of output. This does *not* hold true with wind turbines—each turbine must be connected and regulated separately, and their alternators can't be put in series or parallel.

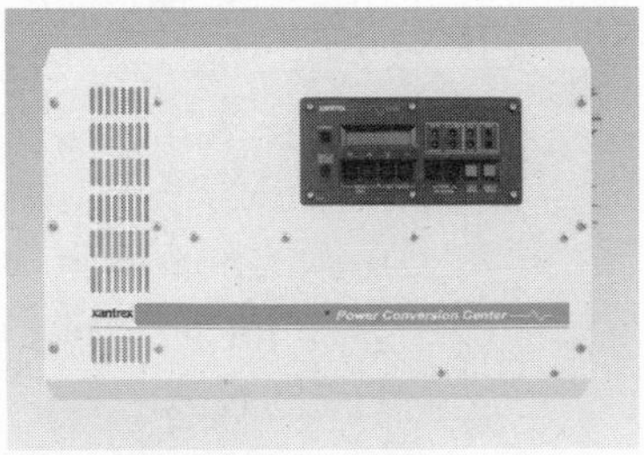

Figure 2.7—Xantrex SW series inverter.
Photo courtesy of Xantrex Technology Inc.

Inverters

Inverters are a common and important component in most RE systems. The inverter in any system converts low voltage DC to 120 VAC for powering normal lights and appliances. Grid-tied inverters also take care of buying and selling electricity to the utility for you, and instantly disconnect from the grid if there's a blackout. Inverters are rated for how many watts they can put out into your loads or into the grid, both continuously and for short periods of time at high power. This "surge" rating is for handling the extra start-up power needs of a large electric motor, for example. It's better to buy a somewhat larger inverter in the first place than to frequently stress out a small one with heavy loads—that can cause future reliability problems.

Figure 2.8—Outback FX series inverter.
Photo courtesy of Outback Power Systems

One area of math that can be confusing to the beginner who is trying to size an inverter and its wiring is again amps, volts and watts. Your typical television set will have a plate on the back that says it draws (for instance), at maximum, 120 watts and 1 amp. But remember that 1 amp in this case is at 120 volts! If your battery bank is 12 volts, the TV will be in reality drawing 10 amps from your battery bank: 120 watts / 12 volts = 10 amps. For a 12 volt inverter that can push out a maximum of 2,000 watts, 2,000 watts / 12 volts = 167 amps from the battery bank. So, both the wires

Figure 2.9—Magnum MS series inverter.
Photo courtesy of Magnum Energy

from the battery bank to your inverter and the wires between each battery in the series and parallel connections must be sized big enough to carry this high current with minimal loss. That means these wires must be big and expensive—and the lower your system voltage, the bigger the wires have to be. Always refer to the manual from your inverter to properly size both its wiring and the battery bank interconnect wiring.

High-tech, expensive inverters such as those offered by Xantrex (Figure 2.7), Outback (Figure 2.8), and Magnum Energy (Figure 2.9) (see our *Sources* chapter for information on finding a dealer) are a joy to own and use. They provide a nearly pure sine wave that is compatible with any and all equipment that runs on electricity, including very sensitive loads. Their surge capacity for starting difficult loads is tremendous, and most include a built-in battery charger and transfer switch—when the backup generator is started, the inverter automatically and seamlessly switches all household loads to run directly from generator power, using any excess to charge the battery bank, and switches everything back to inverter power when the generator shuts down. The switching is done so fast that even computers don't notice the changeover.

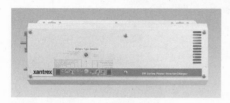

Figure 2.10—Xantrex DR series inverter.
Photo courtesy of Xantrex Technology Inc.

Less expensive inverters are now common, too. They produce a "modified sine wave" output (see the *Meter madness* sidebar in Chapter 20, *Scaling it up and down*), which works fine with almost all electrical equipment. Some sensitive loads, for example certain brands of cordless tool battery chargers and laser printers, will not run off this waveform. Certain light fixtures and electric motors may also make an audible buzzing noise due to the imperfect waveform. Some inverters do not include a built-in battery charger—so that's a component you may have to buy separately. Some manufacturers of top-of-the-line, high-tech inverters also offer smaller modified sine wave versions at a much lower price, too. A typical modified sine wave inverter is shown in Figure 2.10; this one does have a charger built-in. In many cases such inverters are perfectly adequate if sized large enough to handle all of a home's loads. They typically will not have as many "bells and whistles" included, like voltage-controlled relays for turning other devices (such as dump loads or automatic generator start signals) on and off.

Figure 2.11—Typical portable inverter.
Photo courtesy of Xantrex Technology Inc.

Really inexpensive inverters are becoming increasingly common everywhere, too—you can get them at the auto parts store, the truck stop and online for under $200. However, buyer beware! If the inverter is designated as "portable," it's very likely intended to be used only with appliances plugged into the outlets on the front of the inverter (Figure 2.11). If you simply run a cord from one of these plugs to your home main breaker box, the inverter could be ruined in a few weeks, if not instantly, and present a shock hazard. If the inverter lacks a "hard wired" place to connect to a breaker box, it's most likely not intended to be wired to your home. Check with your retailer before purchasing.

Figure 2.12—Commercial-grade battery hydrometer.

Metering

It's essential to know how full your battery bank is to avoid damage from either over or under-charging. Battery bank state of charge can be estimated from the system voltage through experience—but the system voltage varies widely depending on how much energy is coming in or being used. System voltage as an indication of state of charge only becomes accurate when the batteries have had no energy moving into or out of them for an hour or two. The only really accurate way to determine state of charge is by testing each cell of each battery, or a random sample of a few cells, with a commercial-grade battery hydrometer. An example is shown in Figure 2.12. Cheap plastic auto parts store versions with floating plastic balls or swinging levers are not accurate enough. You are looking for a glass-bodied device nearly a foot long (resembling a turkey baster) with a floating glass indicator inside, also calibrated for temperature compensation.

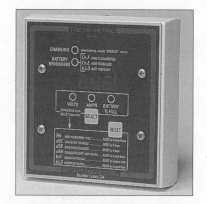

Figure 2.13—TriMetric amp-hour and percent charge meter.
Photo courtesy of Bogart Engineering

Meters are available that count amp-hours and/or watt-hours directly for the entire system by constantly tallying up all energy coming in and being used. These are a wise investment, cost about as much as one battery and are easy to install. They can be connected to track just one component of the system, too—with an amp-hour or watt-hour meter hooked just to your wind turbine, you'll be able to see how many kwh per day, month and year your turbine is producing. Some wind and solar power controllers have this feature built in. However, to track all AC and DC power coming in and out a separate meter (such as the TriMetric, shown in Figure 2.13) must be connected to the system. All power inputs and outputs must go through this device if they are to be measured. As an added advantage, this meter can be set to show your battery bank's calculated state of charge in percent—making it easy for spouses, children and even (possibly) dogs to understand the state of charge, and start conserving energy or turn on the backup generator when stored energy gets low.

Figure 2.14—Xantrex charge controller. Switch-selectable to use for solar power, or as a dump load controller for wind turbines.
Photo courtesy of Xantrex Technology Inc.

Controllers

Because batteries can also be damaged by overcharging, all RE systems include charge controllers (Figure 2.14), even if the control is as easy as a switch that the owner uses to shut down incoming power. Simply turning on extra lights and cranking up Jimi Hendrix on the stereo when the batteries are full and energy is still coming in is also an effective form of control! However, automatic controls are preferable. What if you are not home to hit the off switch on a very sunny and windy day? The actual extra energy coming in doesn't damage the batteries, but does make them "bubble"—they emit flammable hydrogen gas and the electrolyte level drops rapidly. If it drops below the level of the internal plates, the batteries will be damaged.

PV controllers can simply disconnect the panels when the battery bank fills, while wind controllers can't. Most wind turbines must have an electrical load on them at all times and divert any excess to a "dump load"—usually an air or water heating element. If a wind turbine is ever allowed to run unloaded, in high winds it could easily overspeed, run away, and self destruct! Because PV and wind controllers are completely different animals, we'll discuss wind power control later in Chapter 5, *Furling and regulation*.

Shutoffs, breakers and grounding

All power systems incorporate shutoff switches and circuit breakers (or fuses) for various components so that an accidental electric overload won't cause a fire, and so that certain parts of the system can be shut down individually for upgrades and maintenance. Most systems also have a main disconnect switch and main circuit breaker so that *everything* can be instantly shut down with a single switch in case of emergency. Just like with controllers, the shutoffs and breakers needed for wind turbines are completely different from those for the rest of your system, and we cover them in detail in Chapter 18, *Raising*. In short, the breaker that your electrical inspector might require you to install on your wind turbine would be more accurately called a "self destruct switch"—if you turn it off, your turbine is spinning unloaded.

The sample system diagram on the next page (Figure 2.15) is fairly typical. We show buss bars there and haven't mentioned them yet, but they are simply a neat and convenient way to attach the giant tangle of wires coming from your energy sources and going to your loads.

GROUNDING AND LIGHTNING PROTECTION

Grounding is another essential safety precaution in any power system. It reduces the shock hazard and helps protect your system from nearby lightning strikes. The key to properly grounding an RE system is to install only a single central ground rod for the entire electrical system, connected to the main negative battery terminal. Multiple ground rods can cause current flow as a lightning storm approaches, and this can zap both your RE system components and any loads (like TVs, computers and such) that are plugged into your outlets. Wind turbine towers and PV panel frames are usually grounded with separate ground rods right at their location. Tall towers in lightning-prone areas often have a separate ground rod at every guy wire anchor with each guy wire at that location connected to its neighbor, plus another ground rod at the tower base. This does not violate the single ground rod recommendation because here it's only the frames that are grounded multiple times, *not* the electrical wires.

Conservation

We considered listing energy conservation first, as the most essential component of a renewable energy system! But we've already discussed it in the previous chapter, *Introduction to wind power*. Grid power is cheap and heavily subsidized compared to RE. Every dollar you spend on conservation will save you $3 to $5 on the cost of the RE system needed to power all your loads. Enough said!

DOG HAIKU
Napping dogs and cats
Budget energy wisely
Later they can run
Haiku by Michelle Gates

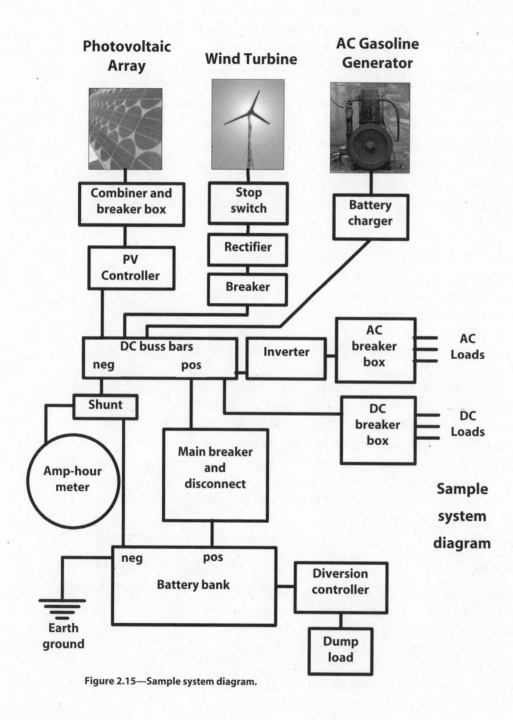

Photovoltaic Array

Wind Turbine

AC Gasoline Generator

Combiner and breaker box

Stop switch

Battery charger

PV Controller

Rectifier

Breaker

DC buss bars

neg   pos

Inverter

AC breaker box

AC Loads

Shunt

DC breaker box

DC Loads

Amp-hour meter

Main breaker and disconnect

Sample system diagram

neg   pos

Battery bank

Diversion controller

Earth ground

Dump load

Figure 2.15—Sample system diagram.

3. Power in the wind

If you are considering building the wind turbine design in this book, it's important to first understand how moving air ends up as electricity. There's a little math involved, but it's quite painless.

A wind turbine extracts energy from moving air by slowing the air down and transferring this harvested energy into a spinning shaft, which then turns a generator to produce electricity.

The difficulty in extracting energy from the wind, compared to other resources like hydro power, solar, or fossil fuels, is that the power available in wind changes rapidly and wildly. Envision a monkey at the controls of a train locomotive and you won't be too far off—there's no way to predict what the monkey will do from second to second. Our fictional monkey could get a wild hair and suddenly jam the "throttle" (the wind speed) to full power, then just as arbitrarily jerk it right back to zero the next moment. We'll come back to the monkey analogy a few times while we explain wind turbine theory.

Work, energy and power

Since we're trying to explain how moving air ends up as electrical energy, we first need to define the terms and measurements used. Keep in mind our first objective: to slow down the wind, and transfer as much of that energy as possible to a spinning shaft. This can be confusing! Work, energy, and power are all very closely related, and it's important to understand why.

Work is defined as a force acting over a distance. It's measured in the metric (SI) unit of joules (j). See the sidebar for an explanation of American units commonly used. One joule equals a force of one newton (n) acting over a distance of one meter (m).

Energy is very closely related to work. The term energy can be used to describe how much work was done (kinetic energy, below), or how much work could be done (potential energy, below). It can also be measured with the same units as work (j). Other units for energy which are important to us here include watt-hours (w-h) and kilowatt-hours (kwh). Energy cannot be created or destroyed, but it can be moved from one system to another—and in the process of doing so, work is always performed.

One common form of energy (and the one we are most concerned with in this chapter) is kinetic energy. Imagine we have a stationary bowling ball in outer space—no gravity and no air. If you could apply a constant force to it over a distance, the ball would accelerate (keep going faster and faster) so long as that force is applied. If the force was suddenly removed, then the ball would cease to accelerate, but it would continue to move at a constant velocity. A certain amount of work was done to get it moving, and now the ball contains the same amount of its own kinetic energy that you expended to push it in the first place.

So, energy was transferred from you to the bowling ball. You could have pushed lightly (small force) on the ball over a long distance, or you could have pushed very hard on the ball (lots of force) over a short distance—the same work would have been done, and the ball would have the same kinetic energy. If you could somehow get around to the other side of the bowling ball so that you were in its way and tried to stop it from moving, then some of the bowling ball's kinetic energy would be transferred to you—you would start moving and the ball would deccelerate some.

Another form of energy is potential energy, which can be thought of as stored energy. It could be a rock on a hill waiting to roll down, a compressed spring waiting for the moment to decompress, or a chemical reaction waiting to happen in your batteries. You can expend a lot of energy and do lots of work rolling a heavy rock up a hill and parking it there. The rock then has stored energy, which would be released into kinetic energy when somebody goes and kicks it back down the hill.

Power is simply the rate at which work was (or can be) performed. Electrical power is usually measured in watts, and mechanical energy is usually measured in horsepower. One horsepower equals about 746 watts. The same amount of work can be done with a tiny bit of power, or lots of it—so what matters is how long it took to perform that work.

Wind turbines are basically a long, convoluted way to gather solar energy. The sun heats the earth in different places, the density of gigantic air masses changes, and then the air moves around—it's being pushed by the sun! In other words, thermal energy from the sun is transferred to the atmosphere, giving it kinetic energy. Our goal with a wind turbine is to convert some of that kinetic energy in the wind into mechanical energy at the spinning shaft of the turbine. Just like with the bowling ball in space analogy, this transfer of kinetic energy has the effect of slowing the wind. It is accomplished with spinning turbine blades, as they are the best way to achieve this energy transfer that anyone has yet figured out.

At the end of the day—it all boils down to force over distance and time!

We're still not done converting energy yet, though. The next conversion must be from the mechanical energy in the spinning shaft to electrical energy in the alternator. Once we get into talking about electricity production or use, energy is measured in watt-hours (w-h) or kilowatt-hours (kwh). One kwh equals 1,000 w-h,

A horse, of course

Here in the non-metric United States of America, there are some strange units of measurement still commonly used. One of the oddest is "horsepower" (hp), but it's *not* our fault! Horsepower is used *worldwide* to measure the instantaneous output potential of automobile engines, electric motors, and more. It's a measurement of power, similar to watts. But how did horses get involved in the first place, and how much power can a horse actually produce? Of course the quick answer is, "that depends on how much horse kibble you feed it!" as covered in our discussion of work, energy and power here. But the long answer is interesting, too.

The term "horsepower" was actually a marketing ploy started by steam engine innovator James Watt around 1782 to increase the sales of his improved steam engine. Many coal mines back in those days used ponies for driving large pumps to remove water from the lower levels. He didn't receive royalties from mines using the older Newcomen steam engine, so Watt wanted to sell his new steam engine design to the mining companies that still used ponies. The detailed facts of the story are lost in the mists of history, but many accounts say that Watt measured how much energy that a horse could produce turning a large mill wheel for an hour. He also thought that "ponypower" sounded a bit silly. So, Watt coined the term "horsepower" for his marketing plan to mine owners, so they could easily understand their cost savings if they switched to steam power for their pumps.

Here are some other horsepower conversions:

• 1 horsepower = 746 watts

• 1 horsepower = 550 foot-pounds per second

• 1 horsepower = 33,000 foot-pounds per minute

• 1 horsepower = 42.44 BTU (British Thermal Units) per minute

• 1 horsepower = 0.7456999 × kilowatts

• 1 kilowatt = 1.34102 × horsepower

and one kwh also equals 3,600,000 joules. These units are interchangeable. You can use either joules or kwh to measure the total energy produced or used by *anything* over time—your dog shoving chew toys across the sidewalk, lifting a rock over your head, the energy production of your wind turbine, or how much energy it took to watch stupid TV shows all night. The *power* that's being produced or consumed at any given instant by all of these methods is measured in watts, but that's a less important measurement for a renewable energy system.

Energy (in watt-hours) is also what you really need to keep track of! It's not very relevant that your wind turbine was making 300 watts of power the last time you looked at the meter, or that your TV draws 100 watts. The important concepts are: for how many hours did your wind turbine make that 300 watts, and for how many hours did you have that 100 watt TV on?

EFFICIENCY

Remember that our objective here is to slow down the wind and transfer as much of that energy to a spinning shaft as possible. Why a spinning shaft? It's simply the most efficient and convenient way to move this energy to a generator for conversion to electricity.

Why is efficiency important? Because it's a measure of how much of the energy you put into a system is available for you to use at the other end. The laws of thermodynamics state that energy cannot be created or destroyed, it can only be transferred about. They also state that no system can be 100 percent efficient or better—so, to get 100 watts of power output, you have to push the generator with *more* than 100 watts of power input. This explains why you can't mount a wind turbine on your car to increase your gas mileage—see the sidebar on the facing page for a more detailed explanation.

Scam artists and misguided enthusiasts claim to have built machines that are more than 100 percent efficient, and thus get their energy for "free"—these are commonly called perpetual motion machines, free energy generators, and overunity devices. Folks have been trying to build them since the 13th century, and none of them has ever worked. To quote the late Robert Heinlein, "TANSTAAFL: There ain't no such thing as a free lunch." All your energy has to come from somewhere, so it's never "free!"

Vehicle-mounted wind turbines?

People often look at us with stunned incredulity when they see us testing a new wind turbine design, mounted on the front of a vehicle moving up and down our remote road. Actually, we only use this test rarely—it's not a good indication of turbine output, just a quick way to see if a new design is even close to working effectively before mounting it high on a tower top. And it's somewhat dangerous! Shouted comments like "You'll never get it off the ground, you don't have any wings!" add to the general hilarity. Some folks often get a mistaken impression of what's really going on, though. We're testing turbine performance, not trying to move the vehicle with wind power.

No, it won't fly. Yes, it's dangerous. And no, the wind turbine doesn't power the truck—it's on there for testing purposes. Co-author Dan Bartmann is driving, while Tarmac watches intently for irritable moose to scare away.

We often hear the question, "Why can't you mount a wind turbine on the roof of an electric car and make the car power itself?" This is where the laws of thermodynamics kick in—energy cannot be created or destroyed, and no system is 100 percent efficient. Let's consider what would happen if you mounted a wind turbine on your car:

As your car accelerated, the wind turbine would indeed spin up to optimum speed and start making electricity. It would also produce a very large amount of drag on the car, so that you would be using extra electricity (or gasoline) per mile to push the car against the extra drag. To ensure easy math, let's say that extra drag costs you 100 watt-hours (w-h) per mile.

If the wind turbine and drive system for its output was 100 percent efficient (in other words, if it converted all incoming wind energy into extra driving force for the car), you'd break even, with no increase or decrease in your mileage. But since no system is 100 percent efficient, you can't even break even, much less *increase* your mileage!

In addition any power generation system, including a wind turbine, has many efficiency losses in all parts of the system. With wind turbines, an excellent efficiency to reach is 30 percent—most small, commercial wind turbines can't even perform that well, and certainly not at all wind speeds. So, you'd be getting back at most 30 w-h per mile from the turbine, while your vehicle is expending an extra 100 w-h per mile. This is a net loss in mileage, not a gain—you would be *losing* at least 70 w-h per mile in extra electricity or gas by mounting a wind turbine on the front of your vehicle!

Wind power only works when the energy source is free—wind from nature blowing across your property. If you have to generate the wind yourself, it's always a losing proposition.

Power available in the wind

The power in the wind that's available for harvest depends on the wind speed, air density and the area that's swept by the turbine blades. Air weighs roughly 1.23 kilograms per cubic meter at sea level, and this weight is what allows it to perform work on our turbine blades to make power. The formula for power available in the wind is the most important math in wind power technology! Here it is:

Power in watts = 1/2 × air density × swept area × wind velocity³

Where:

- Air density (rho) = 1.23 kilograms per cubic meter (at sea level);

- Swept area is in square meters;

- Wind velocity is in meters per second (1 mph=0.447 meters per second).

There are some very interesting things to note about this formula! First, notice that the wind speed is cubed in there. So when the wind speed doubles, the power available goes up by a factor of eight. To work the formula for a 5-foot diameter turbine in a 10 mph wind, you'd start here:

Radius = 1/2 diameter, so 1/2 of 5 feet is 2-1/2 feet, and 2-1/2 feet = 0.762 meters (m)

The formula for the area of a circle is pi (3.14) × r², so:

Swept area = 3.14 × 0.762² = 1.824 square meters (m²)

Wind speed = 10 mph = 4.470 meters per second (m/s)

And therefore (finally!):

Power available = 1/2 × air density × swept area × wind velocity³

Power available = 1/2 × 1.23 × 1.824 × 4.470³ = 100 watts

Increase the wind speed for this 5-foot rotor to 20 mph (8.941 m/s) and you get:

Power available = 1/2 × 1.23 × 1.824 × 8.941³ = 803 watts

This shows us that there's not much power available in low winds. The only way to increase the available power in low winds is by sweeping a larger area with the blades, and that's the second key concept from this formula. Power available increases by a factor of four when the diameter of the blades doubles. Now work the formula for a 10-foot (3.048 m) diameter rotor (the one you are building from this book) for its 7.297 m² swept area in a 10 mph wind, and you get:

Power available = 1/2 × 1.23 × 7.297 × 4.470³ = 401 watts

And in a 20 mph wind:

Power available = 1/2 × 1.23 × 7.297 × 8.941³ = 3,212 watts

This relationship between swept area, wind speed and power available is exactly why we are building 10-foot diameter turbines instead of smaller ones. Since most wind comes to us at slow speeds (see Chapter 1, *Introduction to wind power*, Figure 1.2), a larger swept area is the only way to get much power from these common wind speeds.

Efficiency losses

Unfortunately, you can't expect your new 10-footer to make 3,212 watts in a 20 mph wind. There are many different factors that eat away at that figure you just calculated for power available in the wind. Losses from the fluid dynamics of moving air, magnetic losses, and electrical losses all conspire to reduce efficiency.

THE BETZ LIMIT

The first big loss comes from the fluid dynamics of moving air. In 1919 Albert Betz calculated that there's a limit to how much power a turbine blade can extract from the wind. As you approach the Betz limit of 59.26 percent energy extraction, more and more air tends to go around the turbine rather than through it. So 59.26 percent is the absolute maximum that can be extracted from the available power in wind. It makes sense—if you extracted 100 percent of the energy the wind would be stopped, and no air could flow into or out of the turbine! The Betz limit is an absolute, and applies to any device that extracts power from the wind.

FRICTION LOSSES

The bearings upon which the wind turbine spins are designed to reduce friction, but it can't be avoided entirely. Friction losses end up as heat inside your bearings instead of the electricity you want to harvest.

Magnetic and electrical losses

There are both electrical and magnetic losses with wind turbine generators when converting that power in the spinning shaft into electricity, and both are wasted as heat. We'll discuss these in detail in Chapter 4, *Electricity from a spinning shaft*.

Coefficient of power (cP)

The final ratio of how much power a wind turbine can extract from the wind to how much power is available in the wind is called the coefficient of power, or cP, of that turbine. It's the technical term for efficiency. Less efficient turbines (lower cP) would need a larger swept area to make the same power from the same wind speed as more efficient turbines (higher cP).

Just by knowing how much power is available in the wind and that the Betz limit applies to anything extracting power from the wind, you are already very well-armed to detect wind turbine scams—and scams are unfortunately very common! First go back to the formula for power available in the wind, and add in cP (which can never be more than 0.5926 thanks to Betz): **1/2 × air density × swept area × wind velocity³ × cP**. If you see a 5-foot diameter wind turbine advertised with claims that it can make 800 watts in a 20 mph wind, you've found yet another scam artist selling a "Betz-beater." Do the math, 800 watts is not possible thanks to Betz.

The "good" and the "perfect" wind turbine?

What cP can you really expect from any wind turbine, including a home-built machine? It's difficult to predict, since the cP will change depending on how much power the turbine is producing at any given time. Wind power researcher Mike Klemen (see the *Sources* chapter for more information) published a website that breaks down cP, wind speed and swept area data for both a "perfect" and a "good" wind turbine. Klemen's fictional "perfect" turbine operates right at the Betz limit no matter what the wind speed is. Giant utility-scale turbines are rapidly approaching this level of cP, but they use expensive and complicated active controls that can't be built in the home workshop. A cP of 35 percent (0.35) from a small wind turbine would be excellent—much better than any small commercial wind turbine out there, at least when averaged across varying wind speeds. Klemen calls that a "good" wind turbine in his data, though no commercial machines have yet reached that cP through all power output levels. If you see an ad blurb for a wind turbine that makes more, look deeper. If yours can achieve a cP of 20–30 percent at most output levels, you did a good job and built it right! See Appendix F, *Useful wind data*, for some interesting turbine efficiency information.

How wind turbine rotors work

Look up "wind turbine" in the encyclopedia or the internet, and you'll be faced with a big array of strange contraptions! In all of them, the "rotor" is defined as the part of the turbine that catches the wind and spins. Some turbines have rotors that spin rapidly, others spin slowly. Some spin parallel to the ground, others perpendicular. Some have two blades, others have 32. In this section, we'll cover the different varieties of wind turbine rotors, how they work, and how and why we selected the design presented in this book.

LIFT VS. DRAG

Designs that use only drag forces to make them spin are the oldest way to harvest wind power, and the easiest to understand. They are often called "drag-based" turbines. Examples are anemometers and various pieces of yard art that are frequently seen spinning in a breeze. The blades or cups push against the wind, and the wind pushes back against them. The resulting rotation is very slow, and the blades or cups that are swinging back around after making power are hurting power output because they are moving the wrong direction: against the wind. The earliest examples of drag-based wind power design are grain grinding and water pumping machines (panemones, Figure 3.1) from Persia and China, with records dating back to 500–1500 AD. Note the wall around the half of the ma-

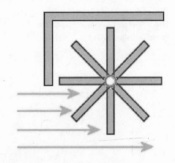

Figure 3.1—Top view of an ancient Persian panemone design for grinding grain or pumping water. The wall can be moved by the operator.
Drawing courtesy of Mike Nixon

chine that is hurting performance by moving against the wind—the operator can move the wall if the wind direction changes to minimize how much wind is blowing into the machine from the wrong direction.

In any drag-based design, the blades can never move faster than the wind. This turns out to be a critical concept for both efficiency and the ease of generating electrical power. The earliest sailboats, such as those used on the Nile by Egyptians as early as 3500 BC, were rigged with square, flat sails. These were drag-based designs, so the boats could only travel in the direction the wind was pushing them, and could never go faster than the wind speed. Fortu-

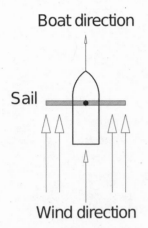

Figure 3.2—A sailboat with a flat sail can only travel with the wind and slower than the wind, never faster.

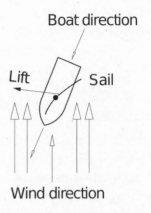

Figure 3.3—A sailboat with a curved, airfoil-shaped sail can travel faster than the wind, and up to 45 degrees against it.

nately for early commerce the prevailing winds there blew them upriver where they needed to go, and the square, drag-based sails (Figure 3.2) served their purpose (and meant a whole lot less rowing by the sailors).

A huge innovation in sailing happened around 200 BC, pioneered by the Arabs. They discovered the concept of aerodynamic lift and started using triangular sails curved into an airfoil shape, just like the cross section of an airplane wing (Figure 3.3). An airfoil shape gives lift, having a curved surface on top and flat surface on the bottom. Air moves over the curved top of the airfoil faster than it does under the flat side on the bottom, which makes a lower pressure area on top that "sucks" the boat along against the wind, or the airplane up to a higher altitude.

Well (harrumph!), that's probably not quite right, though it is the typical "textbook" explanation of lift—and it doesn't explain why airplanes can fly upside down. If you are interested in this complex and arcane physics debate, look up the article *"Lift Doesn't Suck"* by Roger Long, the link can be found on the internet. Instead of sucking, it's easier to simply think of the airfoil as a device that by both its shape and orientation bends the wind to point in the desired direction, with the Newtonian "equal and opposite" reaction propelling the boat with extra force, or holding the airplane up in the air. This added lift force allowed these new ships to travel both up to 45 degrees against the wind and faster than the wind—and does the same thing for a wind turbine blade.

DRAG-BASED WIND TURBINE DESIGNS

The key concept of lift with wind power is that lift forces allow the blade tips of a wind turbine to move faster than the wind is moving. This is important because **doubling the forward speed of the blades quadruples their effect on the wind.** The extra velocity from lift also explains why almost all modern wind turbines are lift-based. All wind turbines (and sailboats) also experience the drag force, but the objective is to minimize drag and maximize lift. There are some drag-based wind turbine designs out there, and they are often the first ones pursued by beginning wind turbine experimenters because the concepts are easier to understand compared to the advantages of lift. Figure 3.4 shows a Savonious rotor wind turbine, and Figure 3.5

Figure 3.5—Anemometer for measuring wind speed, a kind of drag-based wind turbine.

shows a typical anemometer. Both are drag-based. While these designs are not very useful for making electricity because they can never move faster than the wind, they do provide high torque at low RPM from the drag forces. Drag-based designs can be used quite effectively for pumping water or grinding grain—both high-torque operations. But they perform very poorly for making electricity.

Figure 3.4—Drag-based Savonious wind turbine.
Photo courtesy of the AWEA

The wind turbine designs shown at right and on the next page *do* use lift forces to spin, but their drag component is so high that they are still often called drag-based machines. Figure 3.6 shows the Maud Foster mill in Boston, England, built in 1819. It is open to the public for tours and demonstrations—they still mill grain there. It's a Dutch-style windmill. While not exclusively Dutch in origin, these machines were built all over Europe for grinding grain, pumping water, and powering textile and other factories. The Dutch made improvements circa 1390 AD by incorporating a simple airfoil shape onto the backs of the blades, as you can see in the photo. Some machines aligned themselves with the wind mechanically, while others were manually pointed into the wind by the operator.

Figure 3.6—Dutch-style grain-grinding windmill.
Photo courtesy of Ron Fey

Figure 3.7 shows a typical American water pumping windmill. Over six million of these were installed on farms and ranches all over the USA starting in the mid-1800s. They provided mechanical power to drive a pump shaft up and down, and they pointed automatically into the wind via a tail vane. Repair parts and even brand-new wind-

Figure 3.7—American water pumping windmill, lift-based but with very high drag.
Photo courtesy of Dean Bennett Supply

mills are still available. With regular maintenance, these windmills can work hard at pumping water for many decades. The machines are lift-based, but the very high drag in the design makes them problematic for being converted for generating electricity, but still excellent for moving that heavy pump shaft. By the mid-1920s in America, though, a new kind of wind turbine became popular at rural farms and ranches that still had no grid electricity. These new machines were designed to produce electricity—and they relied on strong lift forces, minimized drag forces, and the resulting higher blade speeds to do it.

Companies like Jacobs, Parris-Dunn and WinCharger produced hundreds of thousands of wind turbines for generating electricity, and these machines (Figure 3.8) were common across the American landscape. Their demise started with the great depression and the resulting government initiative (The Rural Electrification Administration, or REA) to extend the electrical grid into rural areas to spur the economy. When grid power arrived, these turbine were no longer needed.

Figure 3.8—A 1930s-era Jacobs wind turbine.

LIFT-BASED WIND TURBINE DESIGNS

Many folks who are new to wind power look at a typical three-bladed turbine and think, "that doesn't look right—too much wind slips between the blades. A whole bunch of wide, flat blades would extract more power from the wind, like an old water pumper windmill." In reality, those three narrow blades are catching much more wind than a multi-blade water-pumper mill ever could because more air is moving through the swept area. It's the same with huge utility-scale wind turbines—the blades may appear to be moving slowly, but they are catching a huge amount of wind. It all goes back to Betz—if you could see the airflow through any wind turbine, you'd find a wake of slow-moving air behind it (slowed down because of the power extracted from it) and a pool of slow-moving air in front. Most importantly, you'd see lots of air going right around the turbine and never being slowed down by it. The multi-blade, high-drag water pumper will have less air going through it and more air going around it than a three-blade, low-drag type.

After this point, we won't discuss drag-based or high-drag wind turbines again. While fun to experiment with for a student's science fair project in the backyard, they are simply not worth the effort to build and raise for making electricity to power a remote home. We'll concentrate on low-drag, lift-based turbines from here on out. There are two basic lift-based turbine designs: the horizontal axis wind turbine (HAWT, Figure 3.9) and the vertical axis wind turbine (VAWT, Figure 3.10).

Figure 3.9—A lift-based HAWT, the one you are building from this book!

HAWT vs. VAWT

HAWTs are what most people envision when they think "wind turbine." The blades spin on an axis parallel to the ground and the entire machine pivots (yaws) around the tower to face the wind. VAWT blades spin on an axis perpendicular to the ground, and they don't need to yaw to face the wind like HAWTs. However, VAWTs are very uncommon, and there have been very few commercially successful models. The reasons will become apparent as you read on. Figure 3.9 shows a typical lift-based HAWT and Figures 3.10 and 3.11 show some different lift-based VAWT designs. Those vertically-oriented blades on the VAWTs pictured have airfoils, and they can rotate faster than the wind speed with relatively low drag.

Figure 3.11—A lift-based VAWT, this one is a Giromill design.
Photo courtesy of the AWEA

Figure 3.10—Lift-based Darrious VAWT.
Photo courtesy of the AWEA

Folks are often attracted to VAWTs because they look so novel and different from "normal" wind turbine designs, and unfortunately this has given rise to a number of wind power investment scams involving VAWTs. Advantages such as no need for a mechanism to yaw the machine into the wind and the (futile, see both below and Chapter 1, *Introduction to wind power*) possibility of locating the turbine near the ground are often cited by VAWT enthusiasts. In reality the designs, physics and performance of VAWTs were thoroughly tested and modeled 30 years ago, and the results of this research (and the engineering problems inherent in VAWTs) are why we don't see many successful commercial VAWTs today.

NOTABLE QUOTES
"Before you try to think 'outside the box,' figure out what's *in* the box and why."
Dan Bartmann, www.Otherpower.com

Unfortunately, the Betz limit applies equally to anything that's flying in the wind, and the effects of turbulence and low wind speeds near ground level do too. No wind turbine, no matter what the design, can perform adequately when flown near the ground where there's little "fuel." All wind turbines require tall towers, putting them at least 30 feet above any obstruction within 300 feet. Anyone that tells you otherwise hasn't done their homework, or is trying to scam you! Don't be surprised if that person soon starts scolding you for not "thinking outside the box." Co-author Dan Bartmann's quote here (above) says it all, in our opinion. Wind power scientists at NREL and Sandia figured out exactly what was "in the box" for VAWTs many years ago, and their experimental results are available for anyone to examine (see the *Sources* chapter). HAWTs are a modern improvement over ancient VAWTs, not the other way around.

Why we don't build VAWTs here

VAWT designs have numerous inherent engineering and reliability problems to overcome, which is why few have ever been commercially successful on any scale. Not all the issues we list apply to all VAWTs, but all are significant. Because half of the machine is not catching the wind (and is actually moving in the wrong direction), the machine must sweep twice as large an area as a HAWT to generate the same amount of power. And since VAWT designs are generally about half as efficient as HAWTs because of the inherently slow blade speeds (Figure 3.14), this amounts to a machine that must be about four times larger than a HAWT to harvest the same amount of power from the wind. Because of the cycle of each blade or cup alternately catching the wind and then moving against the wind, the fatigue stresses on a VAWT are tremendous, so it must be built correspondingly heavier, sturdier and therefore more expensive than a HAWT. The reason most commercial VAWT machines have failed is from breaking apart due to these extra stresses. And since VAWTs don't need to yaw to face the wind, that means they can't yaw *out* of the wind to slow them down! We won't be covering VAWTs again in this book. They can be fun to experiment with, but need to be built very large and at high cost to make any useful amount of electricity.

Wind turbine blade design

All wind turbine blades are a compromise between different factors and forces. Now that we are talking exclusively about lift-based horizontal axis turbines we can concentrate on the most important forces for our objective in this chapter—which is to transfer as much power as possible from slowing down moving air into a spinning shaft. The key concepts to remember here are:

- Lift forces allow the blade tips of a wind turbine to move faster than the wind is moving;

- Doubling the forward speed of the blades quadruples their effect on the wind;

- Both drag and lift forces are operating on any set of blades at all times, and the key to designing a wind turbine blade is to minimize drag and maximize lift.

> **CRUCIAL DEFINITIONS**
>
> **Tip Speed Ratio (TSR):**
> The ratio of how fast a wind turbine's blade tips are moving in their axis of rotation, compared to the speed of the oncoming wind. At a TSR of 1, the blade tips and the wind are moving at the same speed. At a TSR of 6, the blade tips are moving six times faster than the wind—so in a 10 mph wind, the blade tips would be moving at 60 mph.

Tip speed ratio (TSR)

Imagine an unfortunate bug crawling around on the tip of a wind turbine blade. Since the turbine is a horizontal axis lift-based type, the drag forces have been minimized and lift forces maximized so that the tip of the blade on which the bug is riding is moving much faster than the speed of the wind coming at it. The relationship of blade tip speed to the speed of the oncoming wind is called the tip speed ratio (TSR), and the overall efficiency of the blades (how much power they take out of the wind and transfer to a spinning shaft) depends on the TSR. The TSR of a blade depends on multiple factors of its design—if you want to design your own blades, see the *Sources* chapter for some scientific texts (the kind that don't have Dog Haiku in them) that you can research.

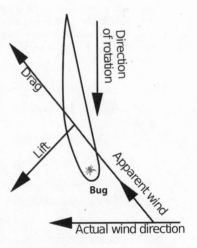

Figure 3.12—Forces on a rotating wind turbine blade, viewed as a cross section from the tip.

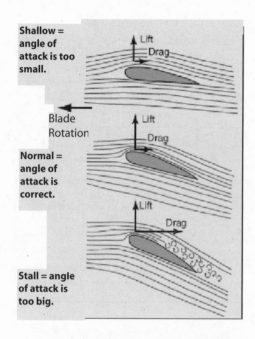

Shallow = angle of attack is too small.

Lift

Drag

Blade Rotation

Normal = angle of attack is correct.

Lift

Drag

Lift

Drag

Stall = angle of attack is too big.

Figure 3.13—The bottom drawing shows a blade tip stalling during start-up: lift is low and drag is high, plus turbulence is produced. The middle drawing shows the tip during normal operation: lift is maximized and drag is minimized. The top drawing shows the tip in very high winds: lift is decreasing because the angle of attack is too shallow.
Drawing courtesy of Mike Nixon

Drag-based machines have a TSR of one or less, which is why we don't use them here. The best TSR range to maximize blade efficiency while keeping tip speeds from getting out of hand (supersonic, for example) is a TSR of 5–6. At a TSR of 6 in a 10 mph wind, the blade tips would be moving at 60 mph.

To our very dizzy bug crawling on the blade tip, the wind feels six times stronger along the direction of rotation than it does from the actual wind direction. In Figure 3.12, let's say the wind is blowing from the right side of the diagram at 10 mph. But since the blade tip where the bug is sitting is moving in the direction of rotation at 60 mph (a TSR of 6), it seems to the bug that the wind is blowing from the lower right of the diagram, from the arrow that's labelled "apparent wind." This arrow represents the real wind speed and direction (10 mph from the top of the drawing) combined with the actual speed and direction of the blade tip and the bug (60 mph from the upper left of the drawing). Apparent wind is the force we are most interested in because it's what the front (leading) edge of the blade, from root to tip, encounters as it moves.

ANGLE OF ATTACK

The "angle of attack" of a wind turbine blade is the angle at which the blade meets the apparent wind. If the angle is wrong drag forces increase, lift forces decrease and the air behind the airfoil becomes turbulent—the blade is said to "stall." Figure 3.13 shows this relationship. The actual wind direction is from the bottom of the diagram, and the lines representing airflow from the left are showing the apparent wind. The blade itself is not actually tilting to change its pitch in this diagram, it's the apparent wind that's changing at different wind speeds—the lowest wind speeds are in the bottom diagram, and the highest wind speeds are in the top diagram.

A blade is not always running at its proper TSR and producing maximum lift. Before the blades start turning in low wind speeds the apparent wind is coming at the flats of the blades, a very poor angle for producing lift. This means they stall, as shown in the bottom diagram in Figure 3.13. Drag forces are high and lift forces are low, so the blade is very inefficient at slowing down the wind.

As the blades gain speed, they start to approach an angle of attack where they can quit stalling and begin to produce far more lift. At this point, a small wind turbine can be visibly observed to pick up speed from the additional lift force, with a substantial increase of power output at the spinning shaft. The blades are now running near their proper TSR. If the blades are moving too fast for the oncoming wind, the angle of attack will be too shallow (the top diagram in Figure 3.13)—once again drag forces increase and lift forces decrease.

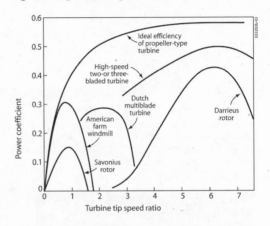

Figure 3.14—Comparison of different wind turbine rotor designs, their tip speed ratio ranges, and their potential coefficients of power (cP).
Graph courtesy of NREL

Different wind turbine designs perform better throughout different ranges of TSR, but a TSR of 5–6 is best for lift-based HAWTs. Drag-based designs lag far behind lift-based models, and VAWTs lag behind HAWTs. Efficiency is less of an issue in wind turbines because the fuel is free—as opposed to with an automobile, where low efficiency equals bad gas mileage. But as we discussed previously, it's still important because an inefficient wind turbine would need to be correspondingly larger, more expensive and heavier than an efficient one to make the same amount of power. Figure 3.14 shows a comparison of potential efficiency in different wind turbine rotor designs. The graph was produced by the wind power scientists at NREL (your tax dollars at work), see the *Sources* chapter for more information. You can see from the graph that no wind turbine design can exceed the Betz limit of a cP of 59.26 percent. The graph also shows how poorly drag-based machines perform compared to lift-based types!

BLADE PITCH AND TWIST

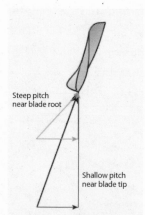

Figure 3.15—Blade twist, shown looking down the blade from the tip.
Drawing by Mike Nixon, courtesy of Hugh Piggott

Figure 3.16—A set of freshly-carved blades, viewed from the tips to show the twist.

Now that your imaginary bug is getting nauseous from riding the blade tip, have him crawl in towards the blade root before he hurls. Wow, all of a sudden he feels much less apparent wind from the direction of rotation! While everything is of course still rotating at the same rpm, the blade roots are moving much more slowly than the tips. This is why most wind turbine blades incorporate a twist into them. The pitch angle is steeper at the root, because that part of the blade is moving slower, and the steeper angle is needed for the correct angle of attack. At the tip there is very little pitch, or none at all, because this is the fastest moving part of the blade and needs less angle of attack to avoid a stall. Refer back to Figure 3.13, and note how the different speeds between blade root and blade tip would affect the lift produced. If a non-twisted blade was running at the "normal" angle of attack shown in the middle drawing of Figure 3.13, the blade root would have an angle of attack that was too large, and the blade tip would have an angle of attack that was too small—the only correct angle of attack would be in the middle of the blade. The twist in a blade keeps the angle of attack into the apparent wind approximately equal from root to tip.

In Figure 3.15, blade twist is shown head on, looking down the blade from the tip. Frankly, extremely simple blades with no twist at all work just fine in a homebuilt turbine. Many successful commercial turbines in the past used straight blades like that. The tips and roots may be stalling at different times, but as a whole the blade is only slightly less efficient. It's entirely feasible to build simple, non-twisted blades for the turbine design in this book. You could carve them with a chainsaw! However, the simple blade design we present for this turbine incorporates a nice simplification of the twist factor that you can easily lay out and produce on any band saw or by hand carving. Since it may be difficult to visualize blade twist from the diagram, you can also take a look at Figure 3.16, which shows the twist along a set of our blades.

> **NOTABLE QUOTES**
> "If it looks like an airfoil—it *is* an airfoil."
> *Hugh Piggott, www.scoraigwind.com*

AIRFOILS

The shape of the airfoil carved into the back of any lift-based blades affects performance even less than the twist. The blade design we use for this turbine is a compromise between best performance and ease of construction. You can download, scale and print airfoil profiles from NASA and spend weeks meticulously carving them, or invest huge amounts of time and money into creating 3D CAD files for your local CNC shop to reproduce the airfoil design to perfection. But at the end of the day, such perfection will gain you at most five or six percent in performance at great expense of money and time. Selecting a likely-looking airfoil profile and carving it by hand also works just fine—the difference in performance compared to scientifically carving a computer-modeled airfoil is small. The simple blade design in this book incorporates twist for optimum lift along the entire blade length at good operating speeds, and the airfoil is quite effective (though it may differ depending on your carving ability). The key is that everything is very easy to lay out and carve by hand, or with power tools. If you want to spend three or four months carving your blades to perfection, that's great. But they'll perform only slightly better than the simplified plan we present in the *Blades* chapter.

NUMBER OF BLADES AND SOLIDITY

We already mentioned that water pumping windmills with many blades work best for high-torque applications like driving a pump shaft, while fewer blades work better for generating electricity. The reason is the fluid dynamics involved in moving air. The more blades, the higher the "solidity" of the rotor as a ratio of swept area to blade surface area. Higher solidity means slower speed and lower TSR, and therefore more torque and slower rpm—plus lower efficiency (Figure 3.14). This is again because the effect a moving blade has on the wind increases with the square of its velocity. And since higher rpm is what we want for electricity generation while high torque is not as important, one, two and three blade designs are preferred. One-blade rotors are impractical, as they are very difficult to design, build and balance.

Two-blade rotors work fine, spin fast, and are quite efficient. But two-blade designs run into a big problem with rotational forces. Envision a figure skater spinning on the ice. When she extends her arms, her rotational speed (rpm) gets slower, because it takes more force to rotate the longer lever of her arms. When she pulls her arms in, her rpm increases. In a two-bladed wind turbine, this change from arms extended (blades out to the side) to arms pulled in (blades vertical) happens twice with every rotation of the rotor! This results in a somewhat violent "chatter" as the turbine yaws to face the changing wind direction. It's enough to be noisy, cause fatigue on all metal and wood turbine parts and possibly vibrate the tail vane right off the turbine—we've been there and done that. A three-bladed rotor gives a good level of solidity and eliminates the chatter. That's why the majority of commercial turbine designs use three blades.

The spinning shaft and the monkey

So, at the end of this long chapter on wind and blade theory we are left with a spinning shaft that is rotating with all the energy we can realistically extract from the wind. The next step is to convert this rotational energy into electricity. The final step will be to put a throttle on that darned monkey to keep him from turning up the wind speed so high that our new wind turbine blows up. Only two more chapters to go on wind power theory!

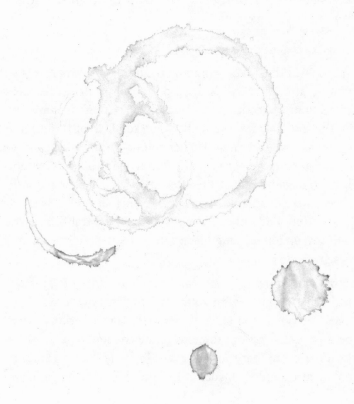

4. Electricity from a spinning shaft

Now we're at the point in this wind turbine theory exercise where we have as much of the available power in the wind as possible being transferred to a spinning shaft, with the deranged, unpredictable monkey still at the controls. The energy in the spinning shaft now needs to be converted into electrical energy for charging our battery bank and then possibly even feeding to the utility grid. The phenomenon of electromagnetic induction is what's used to make this transformation. Move a magnet past a wire, and a voltage is induced. This allows us to transfer the kinetic energy in a spinning shaft into electrical energy.

Magnetic fields

Every magnet is surrounded by a magnetic field, which is stronger closer to the magnet and gets weaker rapidly as you move farther away. The field also has a direction. Figure 4.1 shows the field around a magnet as "field lines," a sort of topographic map of how strong the field is at any distance and which direction its oriented in. The closer the field lines are to each other, the stronger the magnetic field. You can't see this field, but you can see its effects on ferrous metals (those containing iron) by sprinkling iron filings on a sheet of paper on top of a magnet.

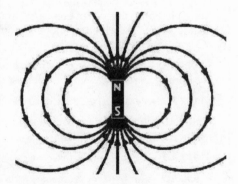

Figure 4.1—Magnetic field lines.

Those field lines might be looking somewhat familiar—just like the magnetic "B-field" around wires that are transferring electric energy, as we discussed in the *What is electricity?* sidebar in Chapter 2, *Renewable energy 101*. For our goal in this chapter of converting the mechanical energy of a spinning shaft to electrical energy at a distance, we have two choices on how to get our magnetic field, electromagnets (the B-field) or permanent magnets.

PERMANENT MAGNETS

Permanent magnets have been known since ancient times. The sidebar later in this chapter, *More than you'll ever need to know about permanent magnets*, goes into great detail about permanent magnets and their properties—for now we're going to stick to only the magnetic theory needed for building wind turbines.

Not all magnets are created equal. Various materials perform better or worse with different aspects of magnet quality, field strength and longevity. Modern permanent magnet formulations have greatly improved over the last few decades—50 years ago they weren't powerful enough to be useful in harvesting the massive amount of energy a wind turbine is capable of extracting from the wind, and electromagnets were used instead. Now that's changed, and in fact the wind turbine in this book gets its magnetic fields from permanent magnets.

Permanent magnet history

• **1200 BC**—Iron smelting discovered, and it's noticed that certain stones (magnetite) attract iron. The magnetite was most likely magnetized by lightning strikes, and was called "lodestone." Lodestones were common in the area of Greece known as Magnesia, hence the name "magnet." The old story that magnets were discovered by a Greek shepherd named Magnes who noticed lodestones sticking to the iron tip of his staff is most likely fiction.

• **1088 AD**—Shen Kuo describes magnetic compass first used for navigation by Chinese. The iron compass needle was magnetized using a lodestone.

• **1187 AD**—Alexander Neckham describes magnetic compass first used for navigation by Europeans.

• **1600 AD**—William Gilbert publishes the book *De Magnete*, which proposes that the Earth is a giant magnet and explains how a compass works.

• **1917 AD**—K. Honda and T. Takai use cobalt with steel to increase the strength of steel magnets.

• **1930 AD**—I. Mishima makes the first AlNiCo (aluminum-nickel-cobalt) magnet.

• **1952 AD**—Scientists from the Phillips Company make the first ceramic (ferrite) magnet, using barium, strontium, and lead-iron oxides.

• **1966 AD**—Dr. Karl J. Strnat of the U.S. Air Force Materials Laboratory produces a $SmCo_5$ (samarium-cobalt) magnet that measures 18 megagauss-oersted (MGOe) in maximum energy product.

• **1972 AD**—Dr. Strnat and Dr. Alden Ray produce a more-powerful (30 MGOe) Sm_2Co_{17} magnet.

• **1983 AD**—General Motors, Sumitomo Special Metals and the Chinese Academy of Sciences simultaneously develop the first NdFeB (neodymium-iron-boron) magnet, at 35 MGOe maximum energy product.

ELECTROMAGNETS

In 1820, Hans Christian Oersted discovered that electricity flowing in a wire created a magnetic field—he noticed a compass needle deflect when he applied electrical current to another of his experiments. If the wire is looped into a coil of multiple turns, the magnetic field strength increases. Increasing the current in the wire also increases the magnetic field strength. When energy flow is stopped, the coil ceases to be magnetic. The advantage of electromagnets over permanent magnets is that the magnetic field strength can be controlled by how much energy is allowed to flow through the coils. With permanent magnets, the only way to control the field strength is to change the distance between the magnet and the wire. This turns out to be a significant factor in trying to generate electricity from moving magnets, and is discussed later in this chapter.

ELECTROMAGNETIC INDUCTION

Michael Faraday discovered in 1831 that a changing magnetic field through a wire induces a voltage in the wire, and this phenomenon is called magnetic induction. It's what makes electric generators and motors possible, and gives us a handy way to take the kinetic energy in a spinning shaft and change it into electrical energy in a wire. The stronger the magnetic

> **NOTABLE QUOTES**
>
> "One day, sir, you may tax it."
>
> *Michael Faraday's reply to William Gladstone, then British Minister of Finance, when asked of the practical value of electricity.*

field and the more rapidly it changes, the more energy can be transferred. The key, though, is the movement. A magnet just sitting next to a wire doesn't make any electricity. Magnetic fields also have a direction, and while this direction is changing from north to south, the direction of the current is also changing.

The conversion from kinetic to electrical energy is not perfectly efficient, either. It always takes more than 1 kilowatt-hour of kinetic energy to produce 1 kilowatt-hour of electrical energy. A magnetic field is not a source of energy, but is simply a very convenient way to transfer it from one form to another. These basic physical facts cause much consternation amongst "over-unity" enthusiasts, and also explain why their perpetual motion machines don't work.

LIBRARY
WAUKESHA COUNTY TECHNICAL COLLEGE
800 MAIN STREET
PEWAUKEE. WI 53072
WITHDRAWN

Generators and alternators

Now we've established that moving magnets next to coils of wire can be used to change kinetic energy from our spinning shaft into electrical energy in a wire. There are many, many ways to design devices that do this! The magnetic field can be provided from either permanent magnets or electromagnets. Either the coils of wire or the magnets can be the moving part while the other sits still. The terminology can be somewhat confusing, too—both generators and alternators are often called "generators" because they generate electricity. Here, we'll separate the two terms: In this book, generators (also called dynamos) make direct current (DC), and alternators make alternating current (AC). These terms were defined in detail in Chapter 2, *Renewable energy 101*.

GENERATORS

DC generators at first seem like a natural for charging the batteries in a remote home. Batteries can charge only with DC, and generators make it, right? The problem is that DC generators are much more complicated to design and build than AC alternators. The electricity flow is changing in direction along with the changing magnetic fields, and that's AC, not DC. So, generators use a device inside called a commutator to keep electric energy flowing in only one direction. This is not something you can easily build at home—it's basically a rotating mechanical switch. Most generators use electromagnets to make the magnetic field for generating electricity. There are also newer DC generators that use permanent magnets instead, but a commutator to keep the power pushing in only one direction for DC is still required.

Figure 4.2—The commutator in this generator is the part towards the front of the picture.
Photo courtesy of Chuck Morrison

Generators were used in automobiles until around 1970, when inexpensive and easy-to-manufacture alternators replaced them. What made this possible was the invention of cheap, tiny semiconductor diodes that convert AC to DC with no moving parts—much easier to build than a commutator. Diodes made the generator obsolete in automobiles.

WITHDRAWN

ALTERNATORS

Alternators are much easier to build at home because they are less complex—the same reason the auto industry uses them now. Like with generators, the magnetic field inside an alternator can be provided by either electromagnets or permanent magnets. Vehicle alternators use spinning electromagnets and stationary coils. We already discussed how the magnetic field strength of electromagnets can be changed by controlling how much electricity flows through them. This makes a car's charging system very simple—when the battery is low, the vehicle's voltage regulator lets more current flow into the alternator's electromagnets, increasing the magnetic field density around them. Thus, the battery charges quickly from the increased alternator power. When the battery is nearly full, the regulator re-

Figure 4.3—A typical vehicle alternator.

duces the current flow into the electromagnets (and thus the magnetic field strength) and energy just trickles into the car battery to keep it topped off. The more electricity a car's alternator is generating, the more physical load is put on the engine—it takes more work to spin the crankshaft, so for example the vehicle uses more fuel when the headlights are on than when they're off. The extra energy to make this extra electricity comes directly out of the gas tank.

Wind turbine alternators

It's easy to make electricity out of the powerful, smooth and constant energy from the spinning shaft of a car's engine. In a wind turbine, we don't have the luxury of smooth power—that darned monkey is still randomly and rapidly switching the wind from high to low power.

Anyone who has lived off the utility grid during a windstorm could appreciate how nice it would be to simply buy a commercial generator or old vehicle alternator from the local hardware store or surplus yard, stick some blades on it, put it up into the wind and generate some electricity. Unfortunately, no such simple solution exists. You can get all kinds of new, used and scrapped alternators and generators in multiple varieties, but none of them are well-suited to generating power from the wind. See the sidebar in this chapter (*Simple surplus solutions*) for more information.

Commercial wind turbine manufacturers design and build their own alternators specifically to match the power curve of their blades. Hardly any of them use DC generators. After trying various commercial and surplus products that were unsuitable, we started building our alternators from scratch too. The trick is matching the output curve of your alternator design to the input curve of power coming in from the blades and spinning shaft. The match is fairly simple to accomplish and can be very efficient when the shaft rpm and torque are constant, as with a gasoline engine or when you can adjust the field strength with a simple regulator, like with a vehicle.

With our erratic and hyperactive monkey at the controls of the wind though, the match is going to be a compromise—we have to decide at what range of rpm and torque input from the spinning shaft the alternator design will be most efficient, and simply accept the losses at other input levels (see Chapter 20, *Scaling it up and down*). For this wind turbine design, we've optimized the alternator to be most efficient at low wind speeds, since they are by far the most common (Chapter 1, *Introduction to wind power*). As winds increase, it gets less and less efficient (more energy wasted as heat), but during those rare high wind events you are already harvesting significant energy, and the extra won't likely be missed very much.

RPM

It's much easier to make electricity from a fast-spinning shaft (high revolutions per minute, or rpm) than from a slow shaft—the output is directly related to the rpm. The low-rpm shaft might be carrying just as much kinetic energy, but more of it will be in torque and less of it in speed. An alternator to make electricity from a low-rpm shaft has to be larger and more expensive than its high-rpm counterpart. Energy is force over distance: A low-rpm shaft has lots of torque (force), and not much distance (it's spinning slowly)—a low-rpm alternator tends to be large in diameter (distance) and contain lots of magnetic material (force).

Simply adding some gearing or belts and pulleys to the spinning shaft seems like an easy way to increase rpm—that's how they get car alternators to spin so much faster than the car's crankshaft, right? The problem is that with wind power we are most interested in the more common low winds where there's not much power available. The friction losses from gearing to increase rpm come right "off the top" in a wind turbine—it won't start making electricity until these losses are overcome, which means a start-up wind speed that's a few mph higher than without the gearing. Ouch! Those low winds are exactly what we want to catch, since high winds are so rare (Chapter 1, Figure 1.1). Gearing would take a major chunk out of our power available in low winds.

PERMANENT MAGNETS VS. ELECTROMAGNETS

Another chunk out of low wind power output that we want to avoid comes from the use of electromagnets to provide the magnetic field. The energy used by the electromagnets also comes first out of turbine output in low winds—the alternator has to be making enough energy to feed these electromagnets before it can start charging your batteries. And, there has to be some way to get electricity to the rotating electromagnets. That's done with devices called brushes and slip rings, both of which are quite difficult to formulate in the home shop. They also wear out and require regular maintenance. Worst of all, if they fail during operation the wind turbine will be free-spinning with nothing to extract the massive amount of energy coming in—that's a dangerous condition that can quickly lead to total failure of the turbine.

THE AXIAL FLUX PERMANENT MAGNET ALTERNATOR

By now we've seemingly brought up a packrat's nest of contradictions about generators. The truth is that any wind generator design is full of trade-offs! An efficient, very low-rpm machine will be heavy, expensive

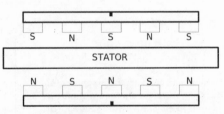

Figure 4.4—Side view of an axial flux air gap permanent magnet alternator.

and difficult to build. A lightweight, inexpensive higher-rpm version will need gearing and/or electromagnets to work with a wind turbine. A good solution, largely developed by Hugh Piggott (see the *Sources* chapter), is the axial flux permanent magnet alternator. The wind turbine blades are attached directly to two spinning steel plates that each have permanent magnets attached, with opposite poles beside one another and opposing one another. These plates (called the magnet rotors) spin on both sides of a flat cast resin disk containing coils of wire (called the stator), "sandwiching" the coils between two discs of spinning magnets (Figure 4.4).

Advantages of this alternator design include:

• Ease of construction: Easy to build from scratch in the home workshop.

• No gearing needed: The magnet rotors are connected directly to the wind turbine blades.

CRUCIAL DEFINITIONS

Start-up speed:
The wind speed at which a wind turbine rotor starts to rotate.

Cut-in speed:
The wind speed at which a wind turbine alternator starts producing electricity. This will generally be a few mph higher than the start-up speed.

• No brushes needed: The magnetic field is provided by permanent magnets.

• Low-rpm operation: For the 10 foot turbine here, power production starts at 140 rpm, with maximum output between 400 and 500 rpm.

• Easy start-up in low winds: Lack of gears, electromagnets and brushes keeps losses to a minimum.

• High efficiency in low winds: And low winds are exactly what we most want to harvest.

Disadvantages of this alternator design include:

• Higher magnet cost: Due to the nature of the design and the low rpm, this machine needs more massive and powerful magnets than more conventional alternators.

• Weight: Low-rpm alternators tend to be heavy.

COMPLETE MAGNETIC CIRCUITS

WHY DON'T YOU...
We get questions like this all the time!

"Why don't you make the magnet rotors out of light weight aluminum or plastic? Seems to me like it would save a lot of weight."

The reason is the complete magnetic circuit, described on this page. Without it, your magnetic flux would be dramatically weakened through the stator coils. Magnet rotors made of 1/4 inch thick steel are the minimum that can be used with this alternator to avoid saturation and loss of flux. Magnets are expensive, and thick steel to complete the magnetic circuit lets them perform up to their potential.

There's a good reason that the magnets for this alternator are mounted on thick steel plates. Ferrous metals concentrate and channel magnetic fields, allowing you to keep the fields where you want them (crossing the coils and making electricity) rather than where you don't (interfering with all the metal parts of your wind turbine, and attracting every bottle cap and paper clip in the neighborhood). If you mounted your magnets for this alternator on wooden discs instead of steel, your power output would be less than half of what you get from steel magnet rotors at any given rpm. If you can stick a paper clip to the back of a magnet rotor, the steel is not thick enough for the magnet size you are using—magnetic flux is being wasted out the back of the rotor because the steel is "saturated." It can't absorb any more magnetic flux and just passes the surplus along to the other side of the steel plate.

AIR GAPS AND LAMINATE CORES

Compared to most conventional alternators, this one has a very large distance between the magnets, and there is no good conductor of magnetic flux between the magnet rotors—that's why this is often called an "air gap alternator." More conventional alternators utilize steel cores inside and around the stator coils to more effectively channel the magnetic field through the coils. These steel cores are always made from thin strips of laminated steel to avoid "eddy current losses" (see next section).

One reason we don't use laminated steel cores in the axial flux alternator featured in this book is to simplify construction. They are difficult and time-consuming to fabricate, and the thin sheets of specially formulated steel needed can be expensive. Each thin layer of high-silicon steel must also be insulated from its neighbor. Before NdFeB magnets were available and affordable, the use of laminated steel cores was the most practical way to get enough flux crossing the coils to make good power at reasonable cost. Now that these magnets are inexpensive, the tedious and expensive extra step of fabricating laminated steel cores is unnecessary. The primitive alternator in this book can be reasonably lightweight, powerful, and efficient with no laminated steel cores by simply throwing a fair bit of very powerful magnetic material at the problem.

Another good reason to avoid laminated steel cores is the "cogging problem." Most generators and alternators with permanent magnets and steel cores will have "preferred" resting places, where the magnets align themselves with the slots in the cores. If you turn such a machine by hand, it will feel "lumpy"—the torque required to turn it varies depending on the position of the magnets in relation to the stator. This causes some vibration while the machine is running, and it also means that a certain amount of torque is required to get the machine turning in the first place. While there are ways to reduce this, it can be a serious problem that can prevent a wind turbine from starting up quickly in low winds. The beauty of the air gap design featured in this book is that cogging is avoided completely, and the required torque to start the rotor spinning must only overcome the friction in the bearings.

The final reason to avoid laminated steel cores is "iron losses." They manifest themselves as heat in the laminated steel cores found in most conventional alternator designs. There are ways to minimize these losses, but they can never be avoided completely in any machine that utilizes laminated steel cores. Magnetic hysteresis is one problem—steel becomes magnetic when exposed to a magnetic field, and wants to retain some of that magnetism. The perfect steel core for use in an alternator could be exposed to a strong magnetic field, and be completely non-magnetic instantly when the field was removed. Unfortunately there is no perfect steel core. The problem of hysteresis is minimized by building steel cores from fairly expensive silicon steel designed for the purpose. Eddy current losses (next section) are the reason that the steel cores used in most electrical devices (motors, alternators, transformers etc.) are laminated.

Fortunately, by avoiding steel cores all together we can avoid all these problems and complications! The only downside is that we need to use a bit more magnetic material to get the same power at any given rpm.

EDDY CURRENTS

Everybody knows that magnets are strongly attracted to ferrous metals. Not everybody has experienced the phenomenon between magnets and any metal that conducts electricity known as "eddy currents." Try moving a magnet across a block of aluminum (for example) and you will notice resistance to movement—the faster you move the magnet, the more resistance you'll feel. Just like in an alternator, the moving magnetic field is inducing electrical currents to flow in the metal, and wherever there is an electric current, there is also a magnetic field induced. In this case the electric current in the metal induces a magnetic field that opposes the moving magnet and resists its movement. You feel resistance for the same reason that an alternator with a load is harder to turn. The trouble is, these eddy currents are wasted as heat, so any eddy currents in an alternator reduce efficiency. Eddy currents can be completely avoided by having no stationary conductors (other than the coils in the stator) near the changing magnetic field.

NUMBER OF PHASES

So now we've explained that an alternator produces alternating current (AC). Alternating current can be provided in one or more phases—most homes utilize single-phase power, which can be carried on two wires (Figure 4.5). The power comes in the form of a sine wave which fluctuates at a given frequency (60 cycles per second in the US). Single-phase power is easy, since it only requires two conductors for transmission, and it's cheap to increase or decrease the voltage with simple transformers. Single-phase does have drawbacks though—when voltage is at or near zero, then little or no current is flowing. At these times the conductors are not being used efficiently. In an alternator, the torque required to turn it varies with the current, causing vibration.

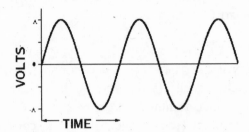

Figure 4.5—Output waveform of one phase of an alternator.

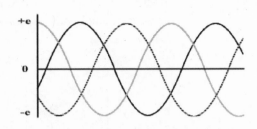

Figure 4.6—Combined output waveform of all three phases in a wind turbine alternator together.

To make better use of the conductors both inside and outside the alternator, and to reduce vibration, multiple phases are often used. Three-phase (Figure 4.6) is the most common, and is the standard for most wind turbines, car alternators, and industrial applications. These three-phase alternators and motors tend to be smaller and more efficient than their single-phase counterparts, and three-phase power can be transmitted on 3 wires. In DC systems (including wind turbines and car alternators), three-phase power is combined at the rectifier to produce fairly smooth DC output.

A simple way to achieve three-phase output from the axial flux alternator described in this book is to use 3 coils in the stator for every 4 poles on a magnet rotor. In the plans here, each magnet rotor has 12 poles (magnets) and the stator has 9 coils. The notes for our experimental 17-foot wind turbine (Chapter 20, *Scaling it up and down*) call for 16 poles and 12 coils. Figures 4.5–4.8 show how this works: at any moment, 3 coils that are 120 degrees apart are "seeing" exactly the same magnetic situation, and therefore those 3

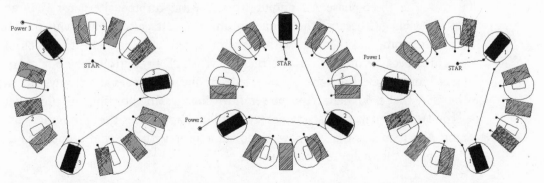

Figures 4.7 (left), 4.8 (center) and 4.9 (right)—These diagrams show how the ratio of 4 magnetic poles to 3 coils produces three-phase AC.

coils are "in phase" with one another. Each phase consists of 3 coils, and the sine wave of each phase is 120 degrees apart from the other two.

Star vs. delta

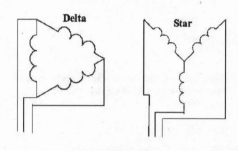

Figure 4.10—A comparison of how delta or star connections are made in a three-phase alternator.

Just like there are two common ways to wire DC circuits (series and parallel) there are also two common ways to wire a three-phase alternator (star and delta). The star connection is also commonly known as a "wye" connection. Note that in Figures 4.7–4.9, one end of each phase is connected to a single point, labeled "star." The other end of the phase goes to that phases power output terminal. The other way to wire up a three-phase alternator is called a "delta" connection. The two different configurations are shown in Figure 4.10. The star connection is a sort of series connection, and it increases the voltage of a phase by connecting it in series with another phase. Delta is more like a parallel connection—so you get only the voltage of a single phase—but resistance is lower and the alternator can produce more current. We always use the star connection in an axial flux alternator. Delta connected alternators, unless they are built to perfection, suffer from parasitic currents flowing between phases which cause some drag and inefficiency in the machine.

Converting wild AC into DC

The three-phase AC output from a wind turbine alternator is often called "wild" AC because the frequency and the voltage vary with with the alternator's rpm, and therefore with the wind speed. It's not useful for much of anything because of that variable frequency, which is controlled by our capricious monkey! Grid-driven AC house current always changes at a steady rate, in the USA that's 60 hertz (the power flow changes in direction 60 times per second). That makes it easy to design and build appliances that use house current. And it means that we have to do something with the wild output of our three-phase alternator to make it usable.

RECTIFIERS

The usual method of converting this three-phase, wild AC wind turbine output into usable form is to send it through a three-phase rectifier. This converts it into DC, which is what we need for battery charging. A three-phase rectifier usually consists of 6 diodes, although it is possible to build one from 3 single phase full wave bridge rectifiers, which each contain 4 diodes. We'll get into more detail on the wiring, schematics, and cost of this in Chapter 13, *Rectifiers*. In short, your three options for rectifiers on this wind turbine are: buy a commercial three-phase rectifier, build your own three-phase rectifier out of 6 diodes, or build your own three-phase rectifier from 3 commercial full-wave bridge rectifiers.

Rectifiers are an inexpensive, simple way to convert AC to DC with no moving parts. Unfortunately, they consume power during operation, which is wasted as heat. In a 12 volt wind turbine, the wasted power is a significant portion of the output, with up to 10 percent loss. In 24 and 48 volt turbines it becomes less of an issue, since the percentage of loss is proportionally smaller. The voltage drop across a three-phase rectifier is usually about 1.4 volts. Using Joule's law, if you multiply the current from the wind turbine by 1.4 you have a good idea how much power is wasted in the rectifier. So, if you have 10 amps coming in then you can bet that losses in the rectifier, wasted as heat, will be about 14 watts. In fact, here we often make some changes in the stators and magnet rotors of 12 volt machines to make up for this rectifier loss. See the sidebar in Chapter 12, *Stator,* for more information on this.

Back to the deranged monkey again

So, now we've taken as much energy from the wind as we could and transferred it to a spinning shaft. We've then taken as much of that mechanical energy as we could and converted it into DC electricity for battery charging. But the monkey still controls the wind speed throttle, and he's cackling insanely with a mean look in his eye!

Let's say he's got the wind moving along at a nice, gentle 15 mph—we're making some nice power from that, about 300 watts from the 1,353 watts available in our 10-foot blade set. A thunderstorm rolls in, and the monkey jams the wind speed throttle up to 60 mph. That gives us 86,600 watts available in the wind, from which our turbine will be attempting to extract about 26,000 watts! The wind turbine will immediately spin up to an extremely high speed and the stator will begin melting, finally bursting into flames—if the blades, frame or tower don't fall apart first. We need a way to put a leash on that monkey so he can't take the wind speed up that high, or a way to reduce how much power the turbine sees at the tower top. And that's what the next chapter, *Furling and regulation*, is all about.

Simple surplus solutions?

We started building wind turbines many years ago the same way that thousands of experimenters around the world try it every year—find an old surplus commercial alternator or generator, stick some blades on it, and try to get it to make some power in the wind! You might get some power this way, it's true. But the result is usually very disappointing power output from a whole lot of money and time invested.

After numerous let-downs and mechanical failures, we started to experiment with building our own alternators. Then we discovered a few folks doing the same thing worldwide, for exactly the same reason--there was nothing out there commercially or surplus that made a good wind turbine alternator. Scottish homebrew wind power guru Hugh Piggott had published plans for his Brakedrum Windmill (see the *Sources* chapter), which incorporated a pickup truck brake drum and wheel bearing for the frame, and ferrite magnets with laminate cores for the permanent magnet alternator. It was heavy and difficult to fabricate, but it worked better for wind power than anything else available at the time. A few years ago we finally got to meet Hugh and attend some of his turbine-building seminars. He had designed an axial flux, air gap wind turbine alternator that took advantage of a recent major price drop in NdFeB magnets. It was lighter, more powerful and much easier to fabricate than the brakedrum design. All of the alternator designs that we present in this book are our offshoots and adaptations of Hugh's original homebrew axial flux alternator plan. All of Hugh's wind turbine books and plans are recommended reading for any homebrew turbine builder!

Here's a quick rundown on what alternators and generators are available commercially and as surplus, and why they don't work very well for wind turbines:

VEHICLE ALTERNATORS

These alternators are cheap and easy to find at any junkyard or auto parts store, but cause major problems in wind power applications. They require about 1,000 rpm to even start making power, and wind turbine blades don't spin nearly that fast. That means gearing is needed to increase rpm, a major power loss, especially in low winds. The current to power the spinning electromagnets is another parasitic loss that really hurts in low winds. The bearings and shaft in a vehicle alternator are not strong enough to support a set of wind turbine blades in high winds, so extra bearings and a bigger shaft are needed—an additional friction loss. Vehicle alternators can be disassembled and re-wound for better wind power performance, but this time-consuming major surgery affects only the rpm at which they start making power (the cut-in speed), and actually hurts higher-speed performance because of increased internal resistance.

AC GENERATORS

The gasoline engine pooped out on the old Honda genset, so why not remove just the AC generator part and attach blades to it? Because to make 120 volt AC power at 60 hz, that engine had to run right at 3600 rpm all the time. A wind turbine generally runs in a range of 150-500 rpm when it's making power, and the rpm varies with wind speed. Getting a constant shaft speed from a wind turbine involves very complex mechanical components—it's usually seen only in large utility-scale turbines.

DC GENERATORS

Back when computer tape drives were the size of washing machines they used big, beefy permanent magnet DC motors to make things spin (Figure 4.11). These motors can still be found as surplus, though they are getting more rare and more expensive. Spin them, and you make electricity. If the motor was originally designed for very low-rpm use and has a voltage rating compatible with your battery bank voltage, a set of blades directly connected to the shaft can perform fairly well for a very small wind turbine. Small is the operative word here, though—these generators can't extract a whole lot of power from the wind. Don't expect more than 75 watts of output, and don't try using anything bigger than a 3-4 foot diameter blade set.

Figure 4.11—Some large, beefy and vintage DC tape drive motors.

The shaft and bearings on these motors are not very strong and are prone to failure, and the brushes need to be checked regularly and replaced when getting thin. If you do try and experiment with a tape drive motor, simply install a big diode in one of the DC output wires to keep the battery bank from just spinning the motor. Try to find a tape drive motor that's rated at as low of an rpm possible (700-900 rpm works well and is relatively common), and with a voltage rating that's about twice your system voltage.

Don't expect to power your remote home with a tape drive motor windmill! It's more along the lines of an advanced science fair project for a high school student.

AC INDUCTION MOTORS AS ALTERNATORS

It's possible to make a three-phase induction motor produce electricity, in either three-phase or single-phase. This requires a controller and capacitor matched to the motor and the load. The generator must run at a fairly constant speed, and our monkey in control of the wind won't ever give us a constant speed. For that reason, this type of generator is far more suitable for constant-speed hydro power or gasoline generator installations than for wind, where shaft speed varies. Using such a motor in a wind turbine would require a complex gearbox to keep shaft speed constant.

VINTAGE WIND TURBINE GENERATORS

Now we're talking! Some of these are highly sought after, and can bring a good price even when they are not working. Most were designed for low-rpm wind turbine use in the 1930s, and there are folks who make a business of buying, rebuilding and reselling them. Old Jacobs generators are especially valuable, and a completely rebuilt and operational vintage Jacobs turbine can fetch US$10,000. If you have a vintage turbine or are considering buying one, carefully look into what you actually have—many pre-REA farmsteads ran on 32 volt battery banks, and you can't buy 32 volt inverters, solar panels or controllers. The wind turbine might work to charge a 24 volt or 48 volt battery bank, thanks to the variable and easily controllable flux in the alternator from the electromagnetic (instead of permanent magnet) field control, but you'll have to do some experimentation or talk to an expert.

More than you'll ever need to know
about permanent magnets!

Since the earth is one giant magnet, compass needles always align themselves to a north–south axis, and the two poles of any magnet are named the north and the south poles. Like poles repel each other, and opposite poles attract. This discovery saved numerous ancient Greeks, Romans and Chinese from being eaten by sea monsters on the open ocean, since they now knew how to point themselves towards their destination instead of sailing in circles.

It's interesting to note, however, that magnet terminology has not yet advanced much in the common vernacular—what is commonly called the north pole of a magnet or compass needle would be better named the "north-seeking" pole. Since opposite magnet poles attract and we refer to the earth's magnetic and geographic north as "north," the part of a compass needle that actually points north is really the *south pole* of that magnetic needle!

Permanent magnets don't start out magnetized. Once the different elements have been combined into a compound (usually by die pressing and then sintering), they are placed in a magnetizing chamber at the magnet factory. A special piece of machinery then hits them with an extremely intense magnetic field. The better the compound the factory started with, the better the now-permanent magnets hold their magnetic properties and the more powerful they are.

The magnets we use for generating electricity with the wind turbines in this book are made from neodymium, iron and boron, commonly abbreviated as NdFeB. As of 2009, this is the most powerful magnet formulation known. NdFeB magnets can transfer a tremendous amount of energy for their size and mass, give an extremely high field density and are very resistant to demagnetization. Some older magnet formulas (below) can become demagnetized simply from sitting in the earth's magnetic field for a few months, while NdFeB magnets lose only about one percent of their strength every thousand years. NdFeB magnets first became commercially available in 1984, and were extremely expensive. Since then the price has dropped significantly, putting them into the realm of possibility for the home experimenter. However, the set of 24 magnets used in this wind turbine design is still the single most expensive component of the machine!

MAGNET FACTS AND MEASUREMENTS

Magnet strength measurements—The units for measuring the field strength (flux density) of a magnetic field are gauss or tesla. 1 tesla = 10,000 gauss. The earth's magnetic field is on the order of 1 gauss. There are different ways to classify and measure field strength:

• **B (flux density):** This is the measurement (in gauss or tesla) obtained when you use a gaussmeter at the surface of a magnet. The reading is completely dependent on the distance from the surface, the shape of the magnet, the exact location measured, and the thickness of the probe and of the magnet's plating. Steel behind a magnet will also increase the measured B significantly by helping complete the magnetic circuit. B is not a very good way to compare magnets, since it varies so much depending on measurement techniques.

• **Br (residual flux density):** The maximum flux a magnet can produce can be measured only in a closed magnetic circuit. Br figures for magnets are generally provided by the magnet manufacturer. This is one way to compare magnet strength—but keep in mind that a magnet in a closed magnetic circuit is not performing any work on anything except the test equipment.

• **B-H curve:** Also called a hysteresis loop, this graph shows how a magnetic material performs as it is brought to full magnetization (also called saturation), demagnetized, saturated in the opposite direction, then demagnetized again by an external field. The second quadrant of the graph is the most important in actual use. The point where the curve crosses the B axis (the vertical axis) is Br, and the point where it crosses the H axis (the horixontal axis) is Hc (Figure 4.12). The point at which the product of Br and Hc is the greatest is BHmax. Very complicated and expensive equipment is needed to plot a B-H curve, so this information is again provided by the magnet manufacturer.

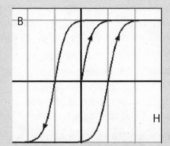

Figure 4.12—A typical B-H curve (hysteresis loop) of a permanent magnet.

• **Maximum energy product (BHmax):** The quality and strength of magnetic materials is best measured by the maximum energy product (BHmax), measured in megagauss oersted (MGOe). This is because the size and shape of a magnet and the material behind it (such as iron) have a large effect on the measured field strength at the surface, as does the exact location at which it measured. In a typical manufacturer's magnet rating (such as N40 or C8) BHmax is the number given, while the letter refers to the magnet formulation (below).

• **Coercivity (Hc):** This measures a magnet's resistance to demagnetization. It is the external magnetic field strength required to magnetize, de-magnetize or re-magnetize a material, also measured in gauss or tesla.

• **Curie Temperature (Tc):** This is the temperature at which a magnet material loses its strength, permanently. Another useful number (if available) is Tmax, the recommended maximum operating temperature. Above Tmax (around 266 degrees F for the NdFeB magnets in this wind turbine design) a magnet will begin to lose its magnetism, and at Tc all magnetic effects are lost.

MAGNET FORMULATIONS

NdFeB (neodymium-iron-boron): This formulation is relatively modern, and first became commercially available in 1984. NdFeB magnets are often called rare-earth magnets. They have the highest B, Br, and BHmax of any magnet formula, and also have very high Hc. They are however very brittle, hard to machine, and sensitive to corrosion and high temperatures. The NdFeB formulation is what we use for the wind turbines in this book, and magnet grades that are acceptable for these turbines range from N35 to N45. At N45 or above, the magnets get very expensive very fast, with only a slight increase in performance. The theoretical maximum grade for NdFeB magnets is N64.

SmCo (samarium-cobalt): Developed in the 1960s and 1970s, these were the first rare earth magnets. They are almost as powerful as NdFeB magnets, and far more powerful than all the others. They are the most expensive magnet formulation, and generally used only when resistance to corrosion and high temperatures are needed. SmCo magnets would work fine in a wind turbine, but don't provide any extra advantages in wind turbine applications over NdFeB to justify the much higher cost.

Ferrite: Also known as hard ceramic, this material is made from strontium or barium ferrite. It was developed in the 1960s as a low-cost and more powerful alternative to AlNiCo and steel magnets. Ferrite magnets are less expensive than NdFeB magnets, but still fairly powerful and resistant to demagnetization. Ferrite magnets are lower in power (B, Br, BHmax) compared to rare earth formulations, and are very brittle. However, they have very high Hc and good Tc, and are quite corrosion-resistant. They were commonly seen in older wind turbine designs, before NdFeB magnets became available at a reasonable cost.

AlNiCo (aluminum-nickel-cobalt): Developed in the 1950s and still in use today. They perform much better than plain steel, but are much weaker in strength (lower B, Br and BHmax) than ceramic or rare-earth formulations, and must be carefully stored since they are prone to demagnetization. Contact with a NdFeB magnet can easily reverse or destroy the field of an AlNiCo magnet.

Bonded (flexible) magnets: a rubberized formulation often seen on refrigerators and magnetic signs. Though they may be manufactured from any magnet formulation when powdered and mixed with rubberizer, the result is always less powerful than a traditional sintered magnet of any formula. Used only where unusual and difficult shapes are needed.

DOG HAIKU

I ate some magnets
Just a puppy, I got stuck
Refrigerator

DOG HAIKU

The shaft spins freely
Extracting power is scary
My tail could get caught

5. Furling and regulation

Here's a summary of our slow progress through wind power theory so far: First we extracted as much energy from the wind as possible by slowing it down with spinning blades, and transferred this energy to a spinning shaft. Then we compromised on a permanent magnet alternator design that has converted as much of that rotational energy as possible into electricity. The problem now is that we still have that deranged monkey at the controls of the locomotive, and he could alternately throttle the wind speed down to nothing, then crank

Figure 5.1—A fully furled wind turbine, protecting itself from high winds.

it right up to 90 mph in a few seconds with no warning. The remaining trick is learning how to make the blades and alternator work together to keep a steady, safe and sane current flowing into the battery bank—even if the wind speed is not at all safe and sane.

Blades + alternator = ?

A big part of experimenting with homebrew wind turbines is trying different ideas and flying them, making changes, and flying them again. The basic math of blade design and alternator design helped us get close, but the empirical approach (trial and error) was how we finally came up with a fairly reliable and easy to build wind turbine. There are many factors in this turbine design that could be changed and make little difference in performance. However, it's critical that the blade and alternator designs be well-matched. Let's look at two extreme cases.

If we took the alternator design from this book and put a 5-foot diameter set of blades on it instead of 10-foot diameter, what would happen? The swept area has decreased from 7.3 square meters to 1.8 square meters, and the power available in a 10 mph wind has decreased from 400 watts to 100 watts. It will now take higher wind speeds to get the entire thing spinning in the first place and then to reach the alternator's cut-in speed, since the smaller blades are attempting to overcome the friction of the bearings. When the alternator does cut in, it will want to convert more mechanical power into electricity at any given rpm than the blades can extract from the wind. This puts a heavy load on the blades and they will not be able to increase their rpm with wind speed, so they will not do a good job of converting the power available from the wind into mechanical energy. The blades will likely stall and rotate slowly, at or near cut-in speed, and the output of the turbine will be dismal. In short, the swept area is too small for the alternator.

On the other extreme, take the 10-foot blade set from this turbine and mount it to a smaller, less powerful alternator designed for a 5-foot diameter blade set. The 5-foot diameter blades turn at approximately twice the rpm of the 10-foot set, so the larger blades on the small alternator would start up in very low winds and spin freely, with no load, up to the cut-in speed of the alternator (upwards of 300 rpm). Past cut-in, the blades would still be in severe overspeed and the alternator would be producing very little power (because it needs to turn even faster). Machines that are overspeeding cannot protect themselves from high winds. There's a good chance the blades could come apart, or the alternator might overheat. So, too small an alternator with too high a cut-in speed will result in overspeeding, noisy blades, dismal power output, and the risk of blade and/or alternator failure.

MATCHING THE BLADES AND ALTERNATOR

A good match between the alternator and the turbine blades is critical. Unfortunately a perfect match is impossible. If you recall from Chapter 1 (*Introduction to wind power*), the power available from the wind is related to the cube of the wind speed. Assuming the blades run at a constant tip speed ratio, then the rpm of the blades is directly related to wind speed. So if you double wind speed, then rpm doubles and the power at the shaft is 8 times as great. Unfortunately the permanent magnet alternator has a linear power curve. So the best we can do is match things up reasonably well over a small range of wind speeds (Chapter 20, *Scaling it up and down*). Usually the goal is to get a pretty good match between about 7 and 25 mph, and then above 25 mph the power from the wind will quickly overcome the power that the alternator can produce. At this point, the turbine should simply try to protect itself while maintaining good output.

Mechanical regulation

When the power at the shaft is greater than what the alternator can convert to electricity, then the blades will overspeed and the alternator will likely overheat. So, we need some system by which we can limit power at the shaft if the wind speed becomes too high. Yes, it's that darned monkey controlling the wind speed again! There are several ways to approach the problem.

MECHANICAL BRAKES

At first glance, some way to simply brake the wind turbine (like the brakes on a vehicle) seems like a logical choice for regulation in high winds. This has been used on some vertical axis machines, but has never been successfully used on a horizontal axis machine that we know of. The trouble with mechanical brakes is that they wear out quickly! In high winds they would have to be applied all the time, converting excess power at the shaft into heat. Just like riding the brakes constantly in your car, the brake gets hot and fails. Mechanical brakes can be effective for *stopping* a wind turbine, but not for regulation.

AIR BRAKES

Some early wind turbine designs from the 1930s used rotational forces to trigger air brakes—big metal scoops that would pivot out and limit the rotor rpm in high winds, then twist back via springs as wind speed dropped. The sidebar on this page shows a vintage example that we were fortunate enough to help spend some time restoring with our neighbor Scotty. How well do air brakes work? They limit the turbine's rpm, for sure—but historically with lots of vibration and noise. Scott's heavily modified version is pretty quiet

Restoring an old beauty

Our neighbor Scott obtained a really neat old wind turbine, a WinCharger from the 1930s. It had a one-piece, two-bladed, 11-foot diameter wooden rotor with copper sheeting stapled to the leading edge for protection. Air brake scoops actuated by centrifugal force deployed to limit the rpm in high wind speeds. The 32 volt DC generator was shot, but he salvaged and repaired the tail and blades. He built a brand new permanent magnet axial flux alternator for it, and let it fly once again. It's been a very reliable and quiet machine for him, and the air brakes work quite well. They make much less noise than he expected.

Figure 5.2—Scotty's old WinCharger, restored by fitting it with a new axial flux permanent magnet alternator.

when the air brakes come out, but we still don't recommend air brakes—they are quite difficult for the hobbyist to design and manufacture, parts will wear out, and ice could jam up the system. If the air brakes fail to deploy due to design or maintenance flaws, the wind turbine will overspeed and could be destroyed.

Figure 5.3—A utility-scale wind turbine, shut down by active controls feathering the blades.

VARIABLE PITCH BLADES (ACTIVE)

This could solve all of our design problems! Change the pitch of the blades dynamically so that the angle of attack (Chapter 3, *Power in the wind*) is always correct for the current wind speed, and you've increased performance substantially in all winds. Plus, you now have a way to reduce power input in high winds—turn them to a really bad angle of attack and they won't produce lift, so the power gathered by the blades is reduced and the turbine is protected. In large utility-scale turbines the blade pitch is continuously adjusted by computers to give optimum power output over a range of wind speeds, and when the wind gets too powerful the blades are turned 90 degrees over to face the wind from their sides instead of the flat face. This effectively shuts the turbine down, and is called "feathering." The big turbine at left in Figure 5.3 has been shut down in this manner. The problem is, the mechanical and electronic systems needed to monitor the turbine and change the blade pitch are extremely complicated and expensive—far out of the realm of what can be accomplished in the typical home workshop and the typical renewable energy budget.

Figure 5.4—The passive, variable pitch blade mechanism on a Jacobs wind turbine.
Photo courtesy of Wind Turbine Industries Corp.

VARIABLE PITCH BLADES (PASSIVE)

There is a somewhat simpler method of incorporating variable-pitch blades, however—change their pitch only for overspeed protection and don't try to match the pitch to the wind speed during normal operation. Some advanced small wind turbine designs use variable pitch blades this way. By "advanced" we don't necessarily mean new or high-tech! Some models from the 1930s used this method very successfully, notably Jacobs designs (Figure 5.4). When wind speed (and thus the rpm) begins to get too high, added weights (or the weight of the blades themselves) force the the blades to twist either towards stall or towards feather, depending on the design. Some very advanced amateurs have experimented with homebrewed passive variable pitch systems, but they are again extremely complicated to design and adjust—out of the realm of possibility for most builders in the home workshop.

REDUCING THE SWEPT AREA

As you learned in Chapter 3, *Power in the wind*, the total swept area of the blades determines how much power can come into the wind turbine. If there was a way to adjust this, our evil wind monkey would be thwarted permanently. Unfortunately, there's no way to change the length of the blades during turbine operation, short of a 12-gauge shotgun—we don't recommend this option, it's messy, loud and problematic! But if there was some way to turn the wind turbine rotor plane of rotation towards the side or the top during high winds, the effective swept area would be reduced. And that's exactly how most small wind turbine designs control themselves. This method has been used in almost all small commercial wind turbines for many years, too. The act of a wind turbine changing alignment to the oncoming wind to reduce the effective swept area is called "furling." Furling systems can be extremely reliable and simple, often using only one moving part.

> **CRUCIAL DEFINITIONS**
> Both of these terms that we use to describe how wind turbines rotate to face either into the wind, or out of it, are nautical in origin.
>
> **Yaw:**
> The movement of a ship around its vertical axis, as in to change heading. In wind turbine terms, yaw is the rotation of the entire wind turbine around the axis of the tower.
>
> **Furl:**
> To roll up or take down a sail on a ship. In wind turbine terms, furling is when the turbine either: 1) yaws to point away from the wind direction, or 2) tilts back vertically. Both serve to reduce the effective swept area of the blades, and are done to protect the turbine from high winds.

Furling systems

There are two axes on which you can turn the blades out of the wind—either up or to the side. There are a few commercial turbines that use tilt-up furling, where in the fully furled, high wind position the blades are spinning parallel to the ground. But some mechanism must be used to keep the machine facing the wind squarely during normal wind conditions, otherwise it will try to tilt back during regular operation. Counterweights, twisted hinges, and springs can be used. All of these are somewhat complicated to build and adjust, and springs are never a good idea on a wind turbine—if they ice up, rust or break, the turbine could be damaged in high winds. Also, since the entire alternator head must tilt back with the blades on it, the output wiring can be exposed to repeated fatigue stresses if care is not taken in the design. All in all, we prefer not to deal with these extra engineering problems since there is a simpler method available—the "furling tail," which yaws the wind turbine to the side when winds get too high. This again reduces the swept area, but now the pivoting, moving part is just the tail, and the entire generator head yaws to follow it. When the tail folds up and in, the turbine follows the tail and yaws to the side. The tail vane always stays parallel to the wind direction.

FURLING TAIL

The furling tail is actually a very old design from waterpumper windmills in the late 1800s. It was perfected for small, homebuilt wind turbines by Scottish wind power guru Hugh Piggott. Variations of the system are also used in most small commercial turbines made today. The turbine frame is designed with a built-in offset—the center of rotation of the blades is not on the same vertical line as the yaw bearing around which the turbine seeks the wind direction (Figures 5.5 and 5.6). The tail is offset at an angle opposite this, and is hinged both upwards and inwards. The weight of the tail holds the turbine in place, pointed straight into the wind. When the wind speed starts to approach the generator's maximum power output capacity, the tail folds both up and in—the thrust against the turbine blades overcomes the weight of the tail and forces the machine to turn to the side. Since the turbine yaws so that the tail vane is always parallel to the wind, the blades end up facing to the side at an angle to the wind, reducing the effective swept area. When wind speed drops into the normal range again, the tail drops back into normal configuration via gravity, and the machine tracks the wind straight on once again.

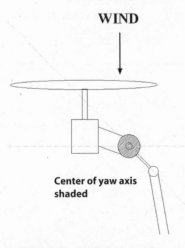

WIND

Center of yaw axis shaded

Figure 5.5—Wind turbine with furling tail in normal operating position, facing into the wind.

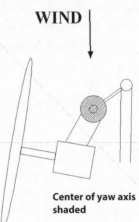

WIND

Center of yaw axis shaded

Figure 5.6—Wind turbine with furling tail in fully furled position, protecting itself from high winds.

Furling tails are nearly foolproof, with no springs or other mechanical components that can ice up, rust or fatigue. Figure 5.7, at right, shows one of our homebrewed wind turbines fully furled, still making near maximum power. It has yawed, and the wind is still moving parallel to the tail. Figures 5.5 and 5.6 on the previous page show how this would look from above. The yaw bearing around which the turbine turns to seek the wind is shaded in the diagrams.

Figure 5.7—Wind turbine in fully furled position, like the top view in Figure 5.6.

Adjusting the Furling Point

The wind speed at which the tail begins to rise depends on a number of factors—the horizontal offset distance between the yaw bearing and the rotor spindle, how far forward of the yaw bearing the blades are mounted, the angle of the tail boom, the size of the tail vane, and the weight of the tail assembly. Most of those factors are permanently set when the frame is welded together, and were determined by trial and error. However, simply adjusting the weight of the tail gives you a wide range to work with. Weight can be easily added, but is hard to remove. A heavier tail assembly lets the turbine spin faster before furling occurs, a lighter one makes it furl earlier. We usually shoot for furling to begin at around 20 mph. If you follow the plans in this book, your furling point will be very close to correct, though you might still want to make small adjustments after watching the turbine in operation.

Furling Tail Design Tips

If you are exploring new territory and planning your own wind turbine design, there are a few rules of thumb you can start with for your first guess at how to design a furling tail system:

• The offset between the main bearing and the yaw bearing should be at least 1/12 the length of a single blade. So, if each blade of your design is 5 feet long (for a 10-foot diameter rotor), the offset should be at least 5 inches.

• The surface area of the tail vane should be between 5 and 10 percent of the swept area. For a 10-foot diameter rotor, that means a tail vane area of between 4 and 8 square feet.

• The tail boom length should be about the same as the length of one blade. That's 5 feet for a 10-foot diameter wind turbine.

• The shape of the tail is not a big factor, and can even be somewhat fanciful. But it's better to keep the tail vane taller than it is wide. Don't bother with any complicated cuts or shapes—while fun to look at, they are prone to fatigue and water penetration.

• To get approximately the correct total tail boom and tail vane weight for this 10-foot turbine, use 1/4 inch thick Baltic birch plywood for the tail. Thicker plywood weighs too much and will mean higher wind speeds before furling begins.

• Don't use sheet metal for the tail vane—there is lots of vibration in the tail boom, and metal is prone to fatigue cracking (Figure 5.8). Plastic could work, but is more brittle than plywood and also tends to crack.

Figure 5.8—Sheet metal tail vane that failed from fatigue cracking.

A leash on the monkey

Now we've put a leash on our theoretical, capricious monkey! Or to stay closer to the analogy, the furling system has put a brick under the throttle so he can't push it down all the way. Figure 5.9 below is a theoretical impression of what power output would look like on a turbine that does not furl. On a 10-footer like the one you will be building from this book, "boom!" would

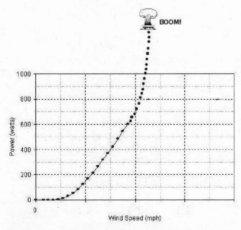

Figure 5.9—Power curve of wind turbines without furling systems in high winds.

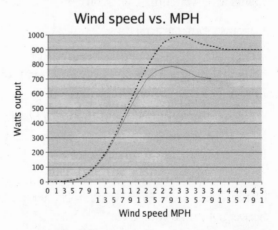

Figure 5.10—Wind turbine power curves that show a furling system kicking in.

probably happen around 35 mph. Figure 5.10 next to it looks much more pleasant. It's an estimated power curve from the turbine design in this book. At the point in the curve where power starts to drop again, the tail is in fully furled position and the turbine is running at an angle to the wind.

Shutdown systems

A turbine with a furling tail can protect itself even in extremely high winds. However, such events are also frequently accompanied by extreme turbulence and rapidly changing wind direction, with violent gusts. Since a furled wind turbine won't be making any more power in 60 mph winds than it was at 30 mph, it doesn't make sense to expose it to this kind of abuse. Earlier and more frequent maintenance will be needed on a turbine that's not shut down during violent winds. Some kind of system to manually shut it down completely is in order.

Mechanical brakes for shutdown are one way to go about it—but they are complicated to build, and prone to failure if not designed and built well. If a brake is your only way to shut the turbine down and the brake system freezes up with ice or rust, or fails in any other way, you have no way to stop the turbine. Systems that manually furl the tail with a cable also work, and can be designed so that if the cable fails, the default position of the turbine is furled out of the wind. Such systems are also tricky to design and build, though. And since wind turbines like to follow the wind on their own, a manual furling shutdown may not stop the wind turbine at all.

Instead, we use an electrical shutdown system for this design. A switch is installed across the three AC turbine output wires, somewhere between the wind turbine and the rectifier. Throwing the switch shorts all three phases together, and the turbine stops within seconds. You'll use this switch to shut everything down during raising and lowering, for protection during extreme winds, and in case something goes badly wrong up there on the tower or in your RE system wiring while the wind is blowing.

If you build your turbine according to the plans in this book, then shorting the alternator will easily stop the machine at any wind speed. Even just shorting a single phase of the three-phase alternator should stop it, so there is a redundant safety factor in this system. If you don't use the alternator design presented here and your own alternator design is not powerful enough, though, it might *not* be able to stop the blades in high winds. This is a common problem in some commercial small wind turbines, but is tolerated because the alternator can be less expensive. We think the fact that this axial flux air gap alternator can be positively and firmly stopped during the most horrific high wind situations is a definite advantage. It's *best* to stop the machine during a lull in the wind when it's already running slowly, because it's hard on the stator and the blades to be stopped quickly in high winds (you may hear an audible "clang")—but we've never damaged a machine here by doing it. The details and schematic for wiring your stop switch are in Chapter 18, *Raising*.

Electrical controls

We still have a problem here, though. Our monkey can't make the wind turbine blow up from too much wind now thanks to the furling tail, but he's also clueless about what's happening at the back of the train, where the battery bank is located. Your batteries are an integral part of your entire wind turbine system—they clamp the wind turbine's high open-circuit voltage down to the system's level, until the batteries are full. At that point system voltage will rapidly climb to levels that can damage your system equipment, and your battery electrolyte will bubble, producing explosive hydrogen gas. Full, overcharging batteries also do not clamp the voltage from the wind turbine down correctly, and the turbine can still overspeed. An overspeeding turbine tends to furl later, so overcharged batteries can actually be quite hard on the machine too. The evil monkey would be happy to pour on the coal all day to ruin the batteries and possibly ruin the wind turbine, so some sort of controller is needed.

Manual controller

The simplest and least expensive controller is manual—if your batteries are full and the wind turbine is still producing power, simply turn on more loads, like lights, stereos, televisions, vacuum cleaners and dishwashers. If you can't consume enough power to keep battery voltage down, just stop the wind turbine entirely with the shutdown switch. Unless you are home all the time, though, it's much safer to have an automatic controller. Wind turbines need to have a load on their alternator at all times, unlike solar panels that can simply be disconnected from the system at any time with a switch.

DIVERSION LOAD CONTROLLER

So, an automatic "diversion load controller" should be installed to divert energy directly from your batteries to a "dump load" when the batteries are full. Usually the dump load is an array of air or water heating elements. Controllers that do this are readily available. The Morningstar TS series (Figure 5.11) or the Xantrex C-series are common choices, and can be converted to run in diversion mode (instead of solar panel mode) with the flip of a switch. Such controllers divert energy directly from the battery bank to the heaters. The direct wind turbine output does not run through them, unlike when used in solar panel control mode. So when used for wind power, the controller unit is the same—but the way you wire it up is completely different. Figure 5.12 shows a block diagram of how a commercial diversion load controller would be wired to your system.

However, once again 12 volt systems get tricky! At 12 volts the machine could easily produce over 80 amps, and neither of those diversion controllers are up to the task—you'd have to use multiple controllers wired in parallel, which is an expensive proposition. With a 48 volt system, you could get by with an inexpensive 40 amp controller. Designing and building your own diversion load heating elements can also be tricky. These loads must be sized large enough to use the entire diversion output of

Figure 5.11—Morningstar TriStar series diversion load controller.
Photo Courtesy of MorningStar Corporation

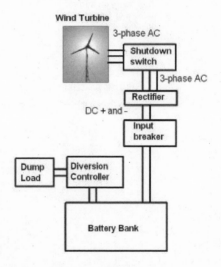

Figure 5.12—Control system wiring digram if using a commercial diversion load controller.

your controller, which itself should be able to divert the entire wind turbine output during high winds. So, we recommend that you size them for 1,500 watts of potential output. The resistance must be exactly right also, to assure that they draw the correct amount of power. We also advise that you simply purchase the correct dump load heating elements from the company that supplied you with your controller, instead of trying to design and build your own—unless you have a very practical and thorough understanding of Ohm's law and circuit design.

CRUCIAL DEFINITIONS

Hysteresis:
A lagging effect, in which the order of previous events can influence the order of subsequent events. If the voltage controlled switch in a diversion load controller did not have hysteresis, the switch would rapidly cycle the dump load heating elements on and off, destroying the high-current relay. By including hysteresis in the controller design, the heaters then cycle on and off only every few minutes.

VOLTAGE CONTROLLED SWITCH AND RELAY

A less expensive alternative to a commercial controller is to use a voltage-controlled switch with hysteresis. Many high-end inverters and PV charge controllers have these built in, often labelled as auxilliary relays (Aux). Instead of diverting a differing amount of power from the batteries to the heating elements as the wind turbine makes more or less power, this simplified system just turns on the heaters full blast until the battery voltage again drops to a reasonable level. The heaters can run on either 120 volt AC house current from the inverter, or DC elements connected with very thick wire directly to the battery bank.

The voltage-controlled switch on a standard inverter or controller can't handle much power. You'll need to use it to turn a much larger relay on or off to start and stop the heaters, just like how the ignition switch and starter relay in an automobile work—the ignition switch you put your car key into switches only a small amount of power, to drive a big relay that allows a whole bunch of power from your battery to flow straight into your starter motor. The relay should be sized to match your dump load.

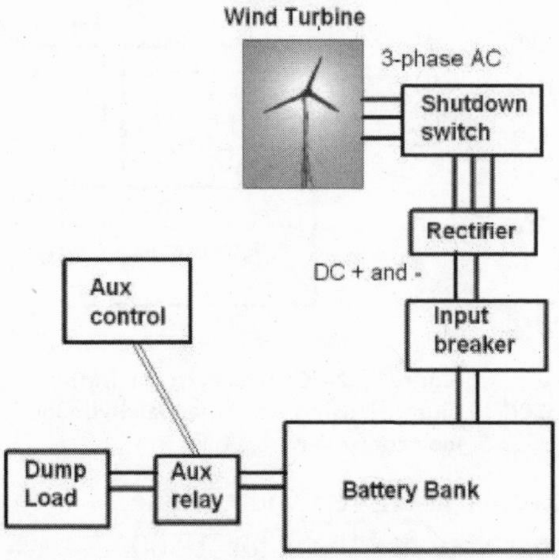

If you have a large and high-tech inverter or PV charge controller, it can be less expensive to use the Aux controller on that equipment to switch 120 volt AC house current to a standard electric heater. But, remember that this dump load should be drawing 1–1.5 kilowatts from your inverter! If the inverter is too small, you won't have any capacity left to run normal household loads, and the inverter will shut down or blow up.

Figure 5.13—Control system wiring diagram is using a voltage controlled switch, such as the Aux relay in an inverter or PV charge controller.

If you do run DC heating elements directly, size the wire appropriately, and use a mercury contactor or solid-state relay—most normal AC relays break down quickly under the high amperage load of a low-voltage DC heating element. DC water heating elements are actually quite easy to find at renewable energy suppliers, so you can easily replace the element in a standard 120 volt AC water heater with a DC version for your dump load. You'll also have to set the correct hysteresis setting so that the contactor doesn't cycle on and off too quickly—the setting will depend on the size and condition of your battery bank. If you go with the Aux relay diversion system, your wind system diagram would look like Figure 5.13.

Watching it all work together

Now we've finally controlled that darned monkey! It doesn't matter what he does to the wind speed, our turbine and its controller can handle anything he throws at us. Once the wind speed gets above 20 mph or so, the wind turbine starts to furl out of the wind, reducing the swept area while continuing to produce lots of energy. If the batteries fill up from this, all the extra energy is sent straight to heating elements while keeping the wind turbine loaded down. If a hurricane or tornado is forecast, the owner can simply shut the machine down manually with the flick of a switch.

Co-authors Dan and Dan enjoy watching wind turbines flying far more than watching television—it's very entertaining and educational! Before building your turbine, see if you can find another small wind installation in your area to watch on a breezy day. Here's what you'll observe, and it serves as a good summary of everything we discussed in these last four chapters on wind power theory:

As the wind speed comes up from zero, the turbine will yaw to face it head-on and the turbine blades will start to rotate at wind speeds of about 5 mph. This is called the turbine's "startup speed." At this point the blades are stalling, because the angle of attack is wrong for the low wind speed. As wind speed increases further, the blades spin faster and the voltage produced by the alternator increases too. When this exceeds the voltage of the battery bank, energy flows into the batteries. This point is called the "cut-in" speed. Somewhere just beyond this point, the wind will reach a high enough speed for the blades' angle of attack to be correct, and the optimum amount of lift is produced. The turbine will visibly speed up, and start producing good power. At this point the turbine is running near its design tip speed ratio (TSR), and it will continue to do so as the wind speed, rpm and power output increase further.

At some point (in this design, around 20 mph winds), the offsetting forces on the blades and furling tail start to diverge and the tail begins to fold up and in. This makes the turbine gradually yaw further out of the wind, following the direction of the tail, keeping the power input from the wind steady instead of letting it increase exponentially with the wind speed. At around 28 mph the tail will be folded part way, and the turbine blades will be pointed away from the wind with power output remaining about the same. The turbine will stay in this position no matter how high the wind gets, with the tail varying in position and the machine yawing to adjust as the high winds vary. As the wind speed starts to drop again, the tail will drop back into normal position via gravity, the turbine will yaw to follow the tail, and the machine will once again be facing the wind head-on.

EVIL MONKEY HAIKU

I control the wind
and smugly laugh at failure
I'm just a primate

I make turbines fail
Engineers try to thwart me
I control the wind

6. Shop safety

This is perhaps the most important chapter of this book! Building wind turbines from scratch is an extremely fun and rewarding hobby, but it involves working with both hand and power tools that can be dangerous. In addition, the finished product and its parts can pose other dangers. Please read this chapter thoroughly. Other chapters will cover these precautions again, plus additional safety issues. Building a wind turbine of this size is not a kid's project, though kids, especially those in high school who are learning wood and metal shop skills, can certainly build a wind turbine *with adult supervision at all times.*

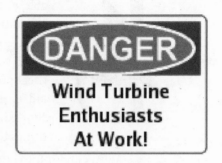

The golden rule of shop safety: THINK!

Think about every possible way the process you are performing could injure you or others. Wear PPE (Personal Protective Equipment) and also make bystanders wear PPE to mitigate these hazards, while being aware that PPE will not completely protect you from all possible hazards. Think about simple, little problems that could cause tragic incidents. Is the power switch for your bandsaw located where someone could accidentally turn it on while leaning against the wall? Could someone trip over that rat's nest of extension cords on your floor? Does that open container of lacquer thinner really need to be perched so precariously on the workbench while you prep parts for casting, or could it be moved to a safer location? Also, try not to work alone if at all possible. Another human may be able to point out any hazards that you didn't notice, and provide a rapid response if an accident happens.

> **NOTABLE QUOTES**
> *"Safety doesn't happen by accident"*
> *Author unknown*

Electrical hazards

Electrical hazards are actually only a minor danger when building a wind turbine, and if you already have a renewable energy system installed, you most likely have learned the precautions. But here they are again, just to be sure! Low-voltage DC power systems do not pose much of a shock danger. At 48 volts, though, you can still get an unpleasant shock. Electrical problems can also pose a fire hazard, not just a shock hazard—a simple car battery can generate enough heat when shorted to start your house on fire. Here are some important things to keep in mind when working with electricity:

• Always check for voltage with a multimeter before touching any wiring, or be sure the equipment is unplugged.

• Frayed 120 VAC power cords for tools and such are a significant shock and fire hazard. Repair or replace all frayed cords. Be careful not to cut your own power cord while using a hand-held power tool.

• Batteries are dangerous in many ways. A wrench dropped across two battery terminals will become red hot in seconds. Undersized wires can heat up enough to cause burns and fires. Battery electrolyte causes chemical burns, and battery fumes are explosive. Batteries are heavy, so lift them properly to avoid back injury.

Figure 6.1—Improper test for voltage potential.

• A wind turbine that is "free-wheeling" with no electrical load attached can generate high enough voltage to be dangerous. If in doubt, check voltage with a multimeter before touching exposed wires. The strange man in Figure 6.1 is not checking for alternator output voltage in the correct manner!

• Use fuses or circuit breakers on all circuits, both low-voltage (12-48 VDC) and high (120 VAC). Carefully read our chapter on wiring your completed wind turbine to your power system, as these safety devices need to be wired using a somewhat different method when protecting a wind turbine.

Metalworking hazards

The most common metalworking hazard is flying metal particles from grinding, cutting and drilling. The tiny metals bits may be moving at high velocity, and they can be both razor sharp and burning hot. Wear a full face shield when grinding or sanding, and at least safety glasses when drilling or cutting metal. Keep bystanders far away or make them wear full face protection, too. Safely store any flammable liquids well away from the grinding area, and be aware that piles of sawdust or a wooden floor could ignite if a hot bit of metal landed on them.

In Figure 6.2, George is demonstrating proper metal grinding procedures. He's wearing a full face shield, his workpiece is securely clamped in a vise, and the power cord for his grinder is placed so it will not be accidentally cut. He is wearing ear protection. The floor is cement, and there is no liquid or solid debris around that could ignite. Some people might argue that he should be wearing welding gloves for this operation, but others would argue that gloves make the tool harder to control. You may receive irritating small burns if you grind without gloves, though. Flying metal particles also stick aggressively to strong magnets, such as those used to build your magnet rotors for this turbine. You'll be in for a tedious session of removing metal particles from your magnets and magnet rotors with duct tape unless you keep them far away from all grinding operations.

Figure 6.2—George demonstrating proper PPE and proper grinding technique.

Grinding wheels can disintegrate without warning during use, so the full face shield is essential. Here are some more metalworking safety issues to keep in mind:

• Grinding, drilling and cutting metal make things hot! If George were to touch that piece of pipe after grinding, he would be seriously burned. Always keep welding gloves around to handle hot metal with, even if you are not actually welding. *Assume every piece of metal in the shop is hot before touching it.*

• Clamp all pieces securely to the worktable when grinding, sanding, planing, cutting, routing, or drilling. Otherwise the tool could accelerate the workpiece to lethal velocities. A very common drill press injury occurs when the operator holds a piece of steel to the table by hand while drilling, instead of clamping it down. Everything goes fine until the bit just breaks through the back of the piece—at that point the bit catches, the workpiece spins, and it shreds the holder's hand into numerous small strips of flesh. Proper setup and clamping take time, but make for a better product and safer working environment.

Figure 6.3—Morwen demonstrates proper PPE.

• Many metalworking operations are loud. Wear ear protection, and make bystanders wear ear protection too. In Figure 6.3, shop dog Morwen the Borzoi demonstrates proper PPE while George is grinding. Morwen must also be careful of getting her long tail caught up in machinery, and her thick, silky coat also poses a fire hazard. Shop dogs must be carefully selected for temperment, behavior, and flammability before being allowed into the workshop while wind turbine construction is in progress! Dogs that don't pay attention to their owners should be banned from the shop. Morwen is a good girl and gets to stay.

• The intense light emissions from arc welding will cause eye damage if the process is viewed without an approved arc welding mask. Keep all bystanders away from the area during welding or make them wear an approved mask. Masks intended for use with a cutting torch are not acceptable for arc welding. Remove shop dogs from the area—if they look too long at the arc, they might eventually need to get a seeing-eye cat due to blindness from retinal burns.

Figure 6.4—Serious safety hazards while welding. George, *what* were you thinking?

• Molten metal particles from welding also present a fire hazard. Be sure no flammable liquids or solids are located near the welding area. In the staged photo at left (Figure 6.4), George appears to have forgotten some very basic welding safety precautions: There's a full gasoline can perched on top of the welder, an open bowl of lacquer thinner and soaked rag right next to his workpiece, and an open lacquer thinner container nearby, too. We won't even elaborate about the empty bourbon bottle there—it's a hazard from both flammability and operator error perspectives.

• The fumes from welding can be toxic. Provide for adequate ventilation and use a respirator if needed. Fumes from galvanized metal are especially nasty—be sure to grind off all the zinc coating before welding galvanized metal.

Magnet hazards

The magnets used in this wind turbine are surprisingly powerful for their size. They are formulated from a mixture of neodymium, iron and boron (NdFeB), and are often called "rare earth" magnets. Most accidents occur when folks are unprepared for how suddenly and powerfully the magnets try to jump out of their hands when brought near another magnet or any ferrous (iron-containing) material. Sharp tools like saws, knives and chisels are especially dangerous around strong magnets. The fellow in Figure 6.5 is moving his box of razor blades far too close to the giant wind turbine magnet rotor—the attraction could easily pull the ferrous razor blades right through the

cardboard box and shred his fingers into human McNuggets®. While the individual magnets used in building this turbine can cause only minor injury (if you follow the safety precautions listed below), a pair of assembled magnet rotors can possibly crush your hand bones into the consistency of banana pulp, and require an assistant with a wooden stick and long lever arm to pry them off your hand so the surgeons at the hospital can go to work. Be sure to store assembled magnet rotors in a very safe location where nobody can accidentally come near them. Here are some more magnet hazards to be aware of:

Figure 6.5—Magnet and razor blade hazard!

• *Always use a strong grip and zen-like concentration when handling large magnets, and wear safety goggles.* Some people wear gloves when handling magnets, this is up to you. Gloves may help prevent accidental blood blisters if the magnets snap together, but they may also compromise your firm grip. Large magnets with sharp corners tend to pinch and shatter the most, rounded ones are safer. Small magnets are generally very safe—but the magnets used here for this wind turbine are very large. Children should be forbidden to handle them, and adults should be duly warned of the dangers.

• *Keep magnets out of reach of children.* Although magnets can be wonderful toys and highly educational, these are *adult* toys. Small children should not be allowed to handle rare earth magnets at all. Older children should handle them only under adult supervision, and wearing proper safety equipment (depending on the size and power of the magnets being used).

• *Magnets can affect medical electronics.* Pacemakers, defibrillators and other medical devices can be sensitive to strong magnetic fields. If you or someone in your household has a pacemaker or defibrillator, or has health issues which require that you wear medical electronics of any sort, avoid magnets completely until consulting with a doctor.

• *Wear eye protection and possibly gloves when handling strong magnets.* They can pose a serious pinch hazard because of their attractive force to each other and to any nearby ferrous object. Usually surprise is an issue—strong magnets can jump out of your hands and snap together from a surprising distance before you realize what is happening. Though small magnets are quite safe, larger sizes are strong enough to pinch you like a pair of pliers, and are very difficult to remove from your fingers without help. They are strong enough that they can fly into each other or to a piece of iron or steel with such velocity that they can break and send sharp fragments flying about! Plus, your expensive magnets are then permanently broken.

• *Magnetic fields can damage magnetic storage media.* These include cassette tapes, floppy disks, credit cards, video tapes and computer hard drives. Keep all strong magnets at least 24 inches away from all types of magnetic media. Don't set your wallet on a strong magnet or your credit cards will not scan at the Quicky-Mart anymore, and the clerk will immediately consider you to be "suspicious."

• *Magnetic fields can damage electronic devices.* Certain gadgets are sensitive to magnetic fields and may be damaged permanently or temporarily disabled if exposed to a strong magnetic field. Be careful with magnets around cell phones, pagers, PDAs, digital cameras and laptop computers. The magnetic field might not affect the storage medium—but it could bend internal metal parts. Any video CRT screen or television will become distorted and/or discolored if a magnet comes too close. While damaged screens can usually be demagnetized, it can be tricky and may require a qualified service technicians to repair, though simply turning the device off and back on again may solve the problem. Store your magnets in a safe place away from electronics of any kind.

Chemical hazards: glues and casting resins

The two most toxic substances used in building wind turbines are casting resin and cyanoacrylate glue (Superglue®). Both have different hazards associated with them. Solvents such as lacquer thinner may also be used to clean parts before assembly. Be aware of both the chemical hazards (breathing the fumes, skin contact) and fire hazard (fumes that can ignite) when using any chemical. Wear rubber gloves and eye protection when using any chemical, and wear the proper respirator for the chemical you are using if you have any doubt as to the safety level of the nasty aroma you are smelling. Dust masks are fine for woodworking, but not for chemicals. Use the exact respirator filter recommended on the can of casting resin that you buy. The local hardware store can set you up with an excellent respirator and the proper filters for around $20.

• Casting resin is very nasty, both the resin part and the hardener. It's best to cast stators and magnet rotors outside, but this can be difficult during cold or rainy weather. At the very least keep windows open and run a window fan for more ventilation. Wear a respirator, eye protection and latex gloves. Have a sink, soap and water available nearby in case of accidental skin contact, and be sure you have enough water available for 10 minutes of flushing in case of eye contact. If it does get in your eyes, call 911 for an ambulance immediately.

• Cyanoacrylate glue and accelerator are also toxic, though the quantities used in this wind turbine are so small that a respirator is not required. But the warnings above about skin and eye contact still apply. Wear latex gloves and eye protection. Call an ambulance if it gets in your eyes. The additional hazards from this glue are the nearly instant bonding of skin and the heat produced by the reaction, especially if accelerator is used. If you accidentally glue your fingers (or worse) together, apply glue remover or acetone and gently and slowly work your fingers apart. The heat from larger quantities of this glue setting all at once is enough to cause serious burns. Co-author DanF recently spilled about 2 ounces of super-thin viscosity cyanoacrylate glue on himself. If this happens, you have only three to four seconds to remove all affected clothing or you will get burned. Once the glue wicks into cloth, the heat produced as it rapidly sets is tremendous. DanF was fortunate to escape both unburned and unobserved, frantically stripping off his clothes.

• We use boiled linseed oil for finishing our wooden rotor blades. It's mostly harmless, but be sure to dispose of oil-soaked rags properly—they can spontaneously combust if left in a pile or thrown in a trash can. Dogs also seem to really like licking up spilled linseed oil, but the industrial wood-finishing grade can contain toxic byproducts. As a nutritional supplement to make your dog's coat more lusterous, buy your linseed (also known as flax seed) oil from the pet store, not the hardware store.

Woodworking hazards

Many of the hazards covered in the metal-working safety section also apply to woodworking. Wear eye and ear protection, a dust mask and a full face shield if needed. Keep bystanders away. In the photos here, George is demonstrating how *not* to use certain pieces of wood and metal working machinery. In Figure 6.6, he got his long hair caught in the drive belt of the lathe! This is very painful, and could possibly cause spinal injury. Tuck in all loose hair and clothing. In Figure 6.7, George got his gold necklace caught in the drill press! Be careful of hanging jewelry, too. A beer in hand is a bad idea too, as shown—not very safe, and your beer gets wood and metal dust in it. Finally, don't cut your fingers off! The injury shown in Figure 6.8 did not happen while building wind turbines, but power tools work quickly. Enough said!

Figure 6.6—Cut your hair, hippie!

Figure 6.7—Remove all jewelry first!

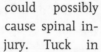

Figure 6.8—Ouch!

DOG HAIKU
I love my master
Power tools have no feelings
I hope he's careful!
Haiku by Michelle Gates

7. Mold building

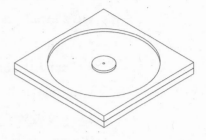

Figure 7.1—Completed magnet rotor mold.

Both the stator and the pair of magnet rotors are cast in resin for strength and protection from the elements. These castings require molds. Like the coil winder, the molds are "tools" that you'll be able to keep and re-use if you build more wind turbines in the future. We use a bandsaw and a homemade circle-cutting jig to fabricate them, so that everything comes out perfectly centered. The molds are also quite easy to build using an electric sabre saw or a hand-powered coping saw, too. Power tools are not needed, they just save you some time.

Building the magnet rotor mold

The turbine uses two magnet rotors that spin with the blades. Each rotor is made from a 12 inch diameter steel disc with 12 magnets on it. Once the rotors are assembled, you'll cast them in resin to keep the magnets in place and prevent corrosion. We use plywood for the molds. Like many other things in these plans, there are alternative ways of doing things—you can use your imagination if you think you've found an easier way. Here we'll detail exactly how we do it. Some of the pictures show two molds in use, but you only need one. We use two molds to save time, because we build lots of wind turbines. Figure 7.1 shows the completed mold, and Figure 7.2 shows the top view, with dimensions. We use Baltic birch plywood for all of our molds because it's of extremely high quality, strong, and has very few splinters.

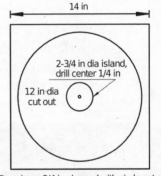

Top piece=3/4 in plywood with circle cut out
Bottom piece=3/4 in plywood

Figure 7.2—Magnet rotor mold dimensions.

Magnet rotor mold materials list:

- 3/4 inch thick plywood, two squares, 14 inch x 14 inch
- 1/2 inch thick plywood, one disc, 6-1/2 inch diameter
- 1/4 inch thick plywood, one disc, 2-3/4 inch diameter
- About 20 wood screws, 1 inch long
- Silicone caulk
- Boiled linseed oil

The mold is 14 inches square. The cutout in the top piece of plywood is 12-1/2 inches in diameter and 3/4 inch deep (because it's cut out from 3/4 inch plywood). In the center is a disc for centering the rotor, 2-3/4 inches in diameter and cut from 1/4 inch thick plywood. The mold lid is the scrap from cutting out the center of the mold, a 12-1/2 inch diameter disc, 3/4 inch thick. The 6-1/2 inch, 1/2 inch thick tapered disc will be used as an "island" during the casting process. The center of every part has a 1/4 inch hole drilled through. This hole is used to center everything during assembly, and it's ex-tremely important that all your cutouts and discs are perfectly centered—if they are off, the magnets will not track correctly over the coils when spinning. All the discs and the inside of the top that you cut out should be sanded smooth, and preferably slightly tapered so that the top of the hole is slightly larger in diameter than the bottom of the hole, so the casting slips out from the mold more easily when set.

Figure 7.3—Magnet rotor mold parts, with mold caulked and ready for casting.

Once all the parts are cut out, screw the two square pieces together so that the one with the 12-1/2 inch diameter tapered hole is on top. Then screw the small 2-3/4 inch diameter disc down in the center, using a 1/4 inch drill bit as a pin to center it perfectly. Use a good number of screws spaced less than four inches apart to assure there are no gaps between the pieces. Wipe the mold down with linseed oil and wait for it to dry, then caulk all the seams inside the mold so that resin can't run into the cracks. Smooth the caulk with your gloved fin-gers so that it covers all interior surfaces and edges of the mold. This, the sanded surfaces and the tapered hole in the mold will make the cast pieces easy to remove after they set up and harden. Your magnet rotor mold is now complete! Set it aside and let the caulking dry overnight.

Building the stator mold

The stator consists of nine coils equally spaced around a circle, wired in three-phase star configuration. Once you finish winding and connecting the coils, you'll need to cast them in resin to form the stator, and this mold (Figure 7.4) is the tool used to get everything lined up perfectly. The stator casting is 15-1/2 inches in diameter, with a hole in the middle 5-3/4 inches in diameter. Once again the mold for this is built from Baltic birch plywood. However, for the

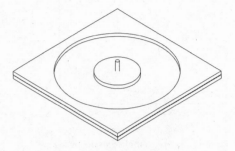

Figure 7.4—Completed stator mold.

stator mold all the pieces are 1/2 inch thick. You could use thicker stock for the lid and the base, but the middle *must* be 1/2 inch thick, since the stator itself must be 1/2 inch thick when it is finished and the middle piece controls the thickness of the casting. The dimensions are shown in Figure 7.5.

Stator mold materials list

- 1/2 inch thick plywood, two squares, 18 inch x 18 inch
- 1/2 inch thick plywood, one disc, 16 inch diameter
- 1/2 inch thick plywood, one disc, 5-3/4 inch diameter
- 1/2 inch–13 tpi threaded rod, 2-1/2 inches long
- 1/2 inch - 13 tpi nut
- 1/2 inch washer
- 3/4 inch long wood screws, quantity 13
- Epoxy
- Silicone caulk
- Boiled linseed oil

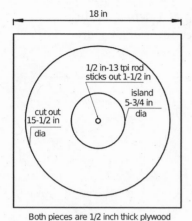

18 in

1/2 in-13 tpi rod
sticks out 1-1/2 in

island
5-3/4 in
dia

cut out
15-1/2 in
dia

Both pieces are 1/2 inch thick plywood

Figure 7.5—Stator mold dimensions.

Start by cutting out an 18 inch square piece of plywood. This will be the bottom of the mold. Find the center of the square. An easy way to do this is to draw light lines from corner to corner—the point where they intersect is the center. Use a compass to draw three circles on the bottom of the mold, as shown in Figure 7.6. One should be 8 inches in diameter, one 12 inches diameter, and one 15-1/2 inches diameter. The space between the 8 inch and the 12 inch circles is the area that the magnets will rotate over, and the coils will need to be centered within these lines. The 15-1/2 inch circle shows the outside diameter of the stator, and this will help to make sure the middle section of the mold is centered perfectly.

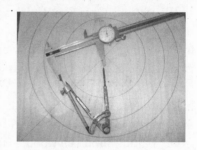

Figure 7.6—Layout of circles on magnet rotor mold base.

There will be nine coils, and it's nice to draw lines on the mold so that you know the maximum permissible size of each coil. It will also help you make sure that the coils are spaced around the circle evenly. In order to divide the circle into nine equal spaces, you need to know the length of one side of the polygon. Work from the 12 inch diameter circle you drew. An easy way is to use the table of sides, angles and sines, Figure 7.7. Multiply the sine of the angle (20 degrees for 9 sides in this stator design, 0.342 from the table) times the diameter of the circle you wish to divide (12 inches). 0.342 × 12 = 4.104 inches per side. This table is reproduced here for the most common numbers of coils we use in these alternators. The angle given is half the actual angle of any segment. Set the compass to the

Polygon number of sides	Angle	Sine
4	45	0.707
5	36	0.588
6	30	0.500
8	22	0.383
9	20	0.342
10	18	0.309
12	15	0.259
15	12	0.208
16	11	0.195
20	9	0.156

Figure 7.7—Table of sides, angles and sines.

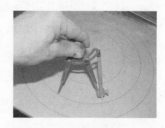

Figure 7.8—Dividing the circle into nine parts with a compass.

length you calculated. Pick a starting point and make marks with the compass about halfway around the circle (Figure 7.8). Then go back to the starting point and work in the opposite direction. This will lessen any error that might come up from imperfect measurement or rounding of numbers.

Next, draw lines from the center to the 15-1/2 inch diameter circle to show the space that each coil should occupy. The finished layout lines should look like those in Figure 7.9.

Figure 7.9—Nine sections of the circle, nicely laid out.

Now make another plywood square just like the last one (18 inches square) and cut out a hole in the center, 15-1/2 inches in diameter. This is the middle layer of the mold, and determines the thickness of the stator casting. Sand the cuts on the inside smooth, and try to give it a very slight taper so that the top is just slightly larger in diameter than the bottom. This will make removing the casting easy. Drill 3/16 inch diameter holes for starting wood screws around the circle, keeping them about 1/2 inch away from the inside edge. You'll need at least nine screws to hold this piece down flat against the bottom of the mold. Run a countersink into the holes deep enough to be sure that all the screws will be flush with (or below) the surface of the mold. This piece, the stator mold base, is shown in Figure 7.10.

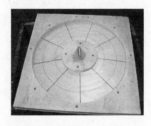

Figure 7.10—Completed stator mold base.

Figure 7.11—Stator mold island.

Cut out another plywood disc 5-3/4 inches in diameter. This will be the island in the center of the mold. It also needs to be sanded smooth and slightly tapered. Drill four holes for wood screws and run the countersink in just like you did for the middle part of the mold. This island piece is shown in Figure 7.11. Now cut out the plywood mold lid, 16 inches diameter, and sand the edges smooth (not shown). Drill a 3/4 inch diameter hole in the center and give it a slight taper with a round file, so that the hole is slightly larger on the bottom (about 7/8 inch). Sandpaper wrapped around a wooden dowel or bolt or a round file work nicely for this tapering operation.

Center the pieces of the mold and screw them together with 3/4 inch long wood screws. If the screws come all the way through the mold, grind the sharp tips off so nobody gets poked by them. Drill a 7/16 inch diameter hole all the way through the center of the mold. Tap the hole to 1/2 inch–13 tpi, and contemplate how nice and easy it is to tap threads into plywood when compared to tapping steel! Put a bit of epoxy on the threads of the threaded rod. You may need to double nut the end of the threaded rod to get a wrench on it, but usually you can turn it in by hand. Screw the threaded rod into the hole so that the bottom of the rod comes flush with the bottom of the mold. If any epoxy comes up to the top of the mold, wipe it off with a rag. There should be 1-1/2 inches of rod protruding from the top of the mold, as shown previously in Figure 7.10.

Wipe the mold down with boiled linseed oil. This will help fill the pores in the wood. It also assures that the mold release grease will not soak into the mold, and the castings will come out of the mold more easily. It also protects the mold from water. Once the linseed oil is dry, run a bead of caulk around the edges of the inside of the mold. This will help create a tapered mold, and it will also plug any cracks between pieces of plywood to prevent resin from seeping in under pressure, to assure the castings come out easily. Put the lid on the mold and hold it there with a 1/2 inch washer and a nut, which will be used later to help clamp the lid down when you are casting. Your stator mold is now done, put it aside until it's time to build the stator!

Figure 7.12—Completed stator mold, with lid attached.

DOG HAIKU
Mold building bores me
Molds do not spin in the wind
I think I'll nap now

8. Coil winder

Before building the actual wind turbine, there are a few tools you will need to make. The coil winder makes it easy to get consistent, tightly-wound coils every time. It's simply a spool made from wood with a hand crank. This chapter will detail exactly how we made ours, but the idea here is simple and there are surely many other ways to build a coil winder. Our design has a fairly large hand crank lever arm to make handling thicker wire easier. The back disc is oversized, with removable steel pins so that both the size and shape of the finished coils can be changed—to build a different alternator design, for example. Our coil winder is also constructed like this because these were the materials we had on hand the day we built it! Your situation may be different, and for a one-time project this could all be simplified some. Before building the coil winder it would be helpful to read Chapter 12, *Stator*, so you can see how you'll be using this tool—that will help you to understand why we make it this way and give you ideas about other possible ways to make a good coil winder. Listed below are all the parts you'll need to build it.

Coil winder parts list

- 1/4 inch thick plywood disc, 6 inch diameter
- 1/4 inch thick plywood disc, 4 inch diameter
- 3/8 inch thick plywood rectangle, 3/4 inch x 1-1/2 inch
- 5/8 inch thick plywood disc, 6 inch diameter
- Wooden dowel, 1 inch diameter, 2-1/2 inches long
- Steel square tubing, 1-1/2 inch, 7 inches long
- 1/4 inch thick steel bar stock, 1 inch by 5 inches, two pieces
- 5/16 inch – 18 tpi bolt, 3-1/2 inches long
- 5/16 inch – 18 tpi bolt, 2-1/2 inches long
- 1/4 inch SAE washer, quantity two
- 5/16 inch – 18 tpi wing nut
- 16d nails, quantity 5

Fabricating the coil winder

Figure 8.1—George demonstrates the use of a hand-held metal-cutting bandsaw to cut out the metal parts.

The first step is to cut out the metal parts. We build lots of wind turbines and find that a portable metal-cutting band saw is fast and easy for this. We use it often, as demonstrated by George in Figure 8.1. An abrasive-wheel chop saw makes it even easier to get your angles and square cuts perfect. If you don't have either one of these saws available, a hack saw, stationary metal-cutting bandsaw, Sawzall® or even a gas cutting torch would work fine. We often have to adapt a design to accommodate the tools and resources that are available to us, and sometimes this requires a bit of imagination! After cutting out the parts, take off all the sharp edges with a hand or bench grinder. If you've never worked metal before, you'll be surprised how easy it is to cut yourself on newly-made metal parts—the burrs can be razor sharp. Figure 8.2 shows the square tubing that supports the coil winder. Cut it off square at 7 inches length, and drill the 5/16 inch hole as shown.

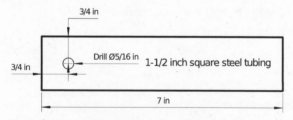

Figure 8.2—Coil winder square tube fabrication.

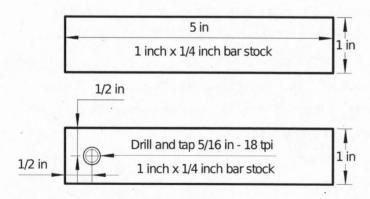

Figure 8.3—Coil winder bar stock fabrication.

Next, cut out the two pieces of 1/4 inch by 1 inch steel bar stock to 5 inches long as shown in Figure 8.3. One piece will be welded to the bottom of the tubing so the machine can be clamped to a workbench. Drill and tap a 5/16 inch – 18 tpi hole in the other as shown. That one will serve as the crank, and the handle will screw into the tapped hole.

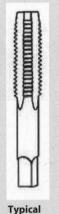

Tips for tapping

In this chapter and many following ones we instruct you to "drill and tap" various holes in metal. Tapping is the process of cutting threads into a hole in the metal to accept a threaded bolt or rod, and the tool used is called a tap. The tap fits into a special tap handle, and the process is usually done by hand. If you don't already own the correct size of tap needed, it's best to buy a tap and drill bit *set* of the correct size and threads per inch (tpi). That's because if the tapped hole needed is (for example) 5/16 inch – 18 tpi, the drill bit required is *not* 5/16 inch! It's instead a size "F." There's a chart of the correct drill bit size for any tap in *Appendix A* of this book.

The tap will have a square end on it that fits into the tap handle. Secure it with the clamping mechanism before starting. Use cutting lubricant when tapping, and be sure the tap is exactly perpendicular to the workpiece. Work slowly, and back the tap out frequently to clear out metal chips. If you twist too hard, the tap will break. Then you have to extract it and buy a new one! This is often called a "tap dance."

Typical tap.

Make the coil winder handle from a 1 inch diameter wooden dowel. Overall it's 2-1/2 inches long. Drill through the center with a 21/64 inch bit so that it turns freely on the 5/16 inch bolt. You will also need to drill in 5/8 inch deep with a 5/8 inch diameter drill bit to accommodate the head of the bolt. We drill it out on a lathe for best precision (Figure 8.4), and while on the lathe it's easy to round the edges and sand it a bit. If you don't have a lathe, you could do this carefully with a hand drill or drill press. If it will be used for just one machine, a perfect wooden handle is not very important, it simply makes for more comfortable winding.

Figure 8.4—Drilling the coil winder handle on a lathe.

Next, take the 5/16 inch, 3-1/2 inch long bolt and drill a 9/64 inch hole through it on center, 2-1/16 inches down from the head, as shown in Figure 8.5. This hole size fits a 16d nail tightly. The nail serves as a cotter pin to lock the spool to the shaft—the bolt *is* the shaft.

2-1/16 in

drill Ø9/64 in

Figure 8.5—Coil winder shaft fabrication.

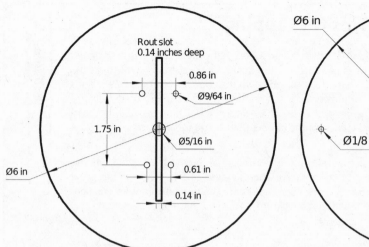

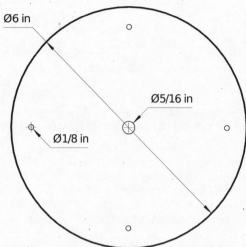

Figure 8.6—Coil winder 6 inch diameter, 5/8 inch thick disc fabrication.

Figure 8.7—Coil winder 6 inch diameter, 1/4 inch thick disc fabrication.

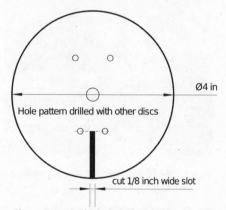

Figure 8.8—Coil winder 4 inch diameter, 1/4 inch thick disc fabrication.

Now you will need to fabricate the three wooden discs. One disc is of 5/8 plywood (we prefer Baltic birch for this) and is 6 inches in diameter, shown in Figure 8.6. The second disc is also 6 inches in diameter, made from 1/4 inch plywood (Figure 8.7). The third disc is smaller, 4 inches in diameter, and made of 1/4 inch plywood. (Figure 8.8). All of these need a 5/16 inch hole in the center. The 6 inch diameter, 5/8 inch thick disc needs a slot routed to accommodate the nail that will be used for a cotter pin. The slot should be 0.14 inch wide, 0.14 inch deep (0.14 is just over 1/8 inch), and 3-1/2 inches long so that a 16d nail with its head cut off fits in there. Both the 6 inch by 5/8 disc and the smaller 4 inch disc need four 9/64 inch holes drilled as shown in Figure 8.6. These holes will accommodate pins around which you'll wind the coils. The 6 inch by 1/4 inch disc needs four 3/32 inch holes drilled so that it does not split out during final assembly. They should be evenly spaced on a 5 inch diameter. Then run a countersink into these holes to keep the screw heads from protruding. Use the wood screws to attach this disc to the thicker one, making sure the heads are flush with the plywood. It's important that the holes line up nicely here. We suggest screwing the smaller disc and the thicker 5/8 inch disc together and drilling all the holes at once.

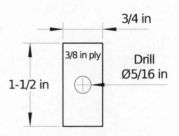

Figure 8.9—Coil winder core fabrication.

To make the coil winder core, cut a piece of 3/8 inch thick plywood as shown at right in Figure 8.9. It's 1-1/2 inches tall and 3/4 inch wide. Drill a 5/16 inch hole through the center. This core serves as a spacer between the discs, and its size and shape determine the final shape of the coils. To wind a different coil size or shape this is the piece you would change, along with the four pins that you'll fabricate later in this chapter.

Weld the 5 inch long bar stock piece that has no holes in it to the bottom (the end furthest from the hole) of the square tubing (Figure 8.10). This serves as a base for clamping the coil winder to a workbench. Notice the grinding marks where George has removed nasty burrs that can cut unsuspecting fingers.

Figure 8.10—Coil winder frame construction welds.

Figure 8.11—Coil winder handle.

Insert the shorter 5/16 inch bolt (2-1/2 inches long) through the wooden handle and thread it into the 5 inch long bar stock that was drilled and tapped to accept the bolt. Put a washer in between the handle and the bar stock. Thread the bolt in so that the threads come flush with the back of the bar stock. The handle should turn freely, and is shown in Figure 8.11 at left.

Now turn the handle over and spot weld where the bolt comes through the bar stock—this assures that the handle will never come loose, and is the leftmost weld in Figure 8.12. Take the longer bolt (the one with the hole drilled in it) and weld it to the other end of the handle. It's the rightmost weld in the same picture, Figure 8.12. The bolt needs to be square with the handle. Put a washer over the bolt that's sticking out (you always want a washer between any part that turns and

Figure 8.12—Coil winder handle welding.

Figure 8.13—Coil winder with 6 inch diameter, 1/4 inch thick disc mounted.

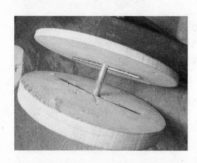

Figure 8.15—How the coil winder shaft lock pin fits.

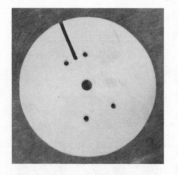

Figure 8.17—Slot cut in 4 inch diameter coil winder disc.

any part that doesn't) and insert the bolt through the hole in the square tubing. Put another washer on the shaft and then install the 6 inch diameter, 1/4 inch thick disc (Figure 8.13). The countersunk holes need to be pointing back, towards the handle. Now insert the locking pin through the shaft and center it, shown at right in Figure 8.14.

Figure 8.14—Coil winder shaft lock pin installed.

Install the 6 inch diameter, 5/8 inch thick disc on the shaft so that it fits over the pin that's through the shaft, and bring it tightly against the 1/4 inch thick disc. The shaft lock pin fits in the groove as shown in Figure 8.15. Use the 5/8 inch long wood screws to attach the two discs together, coming in through the back of the coil winder, through the four holes that you drilled and countersunk earlier. The screw heads need to be flush, they must not stick out.

Put the small rectangular coil winder core on the shaft and align it so that it is centered between all four holes, which will hold the pins around which the coil is formed. Tack it there with a touch of Superglue® so it doesn't rotate. It's shown installed in Figure 8.16 at right. If you have not yet done so, cut the

Figure 8.16—Coil winder core installed.

slot in the front 4 inch diameter disc as shown in the fabrication plan for that disc (Figure 8.9, previous page, and Figure 8.17, this page). You'll start winding the coils from this slot, and it needs to be centered at the narrowest part of the coil.

Now take the four 16d nails and cut them off at 1-1/8 inches long (including the head), as shown at right in Figure 8.18. De-burr the ends you've cut so that they are not sharp. Once the coil winder is finished, you'll insert these pins through the four holes in the front (4 inch diameter) disc and into the holes in the rear disc, so that the discs are pinned together and the pins are supported in both the front and in the back. These four pins will determine the shape of the center of the coil. To adapt this coil winder for making coils of different sizes or shapes, you would adjust both the pin spacing here and the coil winder core (Figure 8.9).

Figure 8.18—Coil winder pins cut from nails.

Figure 8.19—Completed coil winder, front view.

The last step is to install the front disc and use the wing nut to hold it in place. The pictures show a normal nut, but a wing nut is easier to take on and off. Your coil winder is now finished, and we'll discuss how to use it when you start making the stator. The front view is shown at left in Figure 8.19, and the back (handle side) view at right in Figure 8.20.

Figure 8.20—Completed coil winder, back view.

Hand tools vs. power tools

We often hear the question "Can this wind turbine be built without using power tools?" In short, the answer is *yes it can*—but the fabrication process will be much more time consuming! This is a big issue in developing countries and indeed any remote area. Wood is easy to cut, carve, drill and fasten by hand. Building molds and carving blades with hand tools takes a bit more time than with the power versions of those tools, but the gentle scrape of drawknives, coping saws and sanding blocks is much more pleasing to the ear than the obnoxious whine of power planers and electric sanders. When it comes to working metal, though, hand tools are extremely fatiguing and take lots of extra time. An electric drill and electric metal-cutting saw will save you days of toil. And, there's no substitute for welding to make strong metal parts! If you don't have these power tools available, consider hiring someone to do your metalworking and welding for you.

A solar and wind powered caboose

Our good friend Dave M made the big move from town to the mountains a few years back, and joined our tiny off-grid community. He purchased a beautiful piece of land, with a cabin already on it. The unique thing is that the cabin is an old "captive" caboose! Dave is a high school math teacher and needed a simple power system immediately, or else he would've been grading papers each night by kerosene lantern light—that's really hard on the eyes.

Design considerations:

Dave didn't need much power for his caboose—it measures only 10 by 40 feet, which is a small space to light. All he needed to run for loads were lights, tunes and a computer. The range and refrigerator are propane, and his heat is from a coal/wood stove. This meant he could install a very modest system at a fairly low cost. For such a small system, he didn't need to go through the traditional process of tallying up loads, estimating solar and wind input and plugging all the numbers into a system design spreadsheet. Instead, Dave simply examined some local cabin RE systems that were of similar size, and bought what equipment he could afford.

System equipment (it's a 12 volt system):

- One 100 watt solar panel on a rack that tilts for summer and winter angles
- Four golf cart batteries, 6 volt, wired in series and parallel for 440 amp-hours at 12 volts (880 watt-hours)
- One inexpensive 1,500 watt portable inverter
- One homebuilt 10-foot diameter wind turbine

Results:

Dave is very happy with the small power system in his caboose! It provides all the power he can use, and he frequently has to shut the wind turbine down when his batteries can't hold any more power. He doesn't have a dump load controller for the wind turbine, he manually pulls the shutdown switch to stop incoming power. He doesn't even own a gasoline backup generator for emergency battery charging, and has never needed one.

Wind turbine details:

Dave's machine uses a surplus Volvo front strut assembly as the frame, and two Volvo disk brake rotors as the magnet rotors. This design using the car parts was featured in issues 74–78 of *Back Home Magazine* (see Chapter 21, *Sources*), and still makes for a very functional (though somewhat ugly) wind turbine. The newer design featured in this book incorporates many improvements over the original surplus car parts machines like Dave's. Construction time with the new turbines is a day or two shorter since the surplus car parts don't have to be removed, disassembled, cleaned and modified. However, Dave's old turbine is still up and running five years after installation, with only minor, regularly-scheduled maintenance needed!

9. Frame

This chapter will describe how to build the frame of the wind turbine. This assembly includes some very important parts all welded together: the yaw bearing on which the turbine rotates into and out of the wind, the main bearing on which the blades rotate, the pivot on which the tail swings in and out and the mounting bracket for the alternator. It involves a fair bit of metal work. To build the frame you'll need to have the tools and ability to cut, grind and weld steel. If you're new to this type of work it might pay to practice your skills on some scraps. The tolerances and angles for the frame are not terribly important—the design is very forgiving. But some of the welds have a great deal of force put on them during turbine operation, so you need to be a bit careful to insure the quality and strength of all the welds. Be sure the welder you are using is powerful enough to penetrate 1/4 inch steel, and be sure you use the correct type of rod or wire recommended by the manufacturer. A welder that's not powerful enough will leave weak spots inside the welds that you can't see, and a splattery, unattractive exterior. We call these "bird crap" welds because that's what they look like!

NOTABLE QUOTES

"Bits fall off!"

*Hugh Piggott, Scoraig Wind Electric
(Referring to what happens to every wind turbine—eventually.)*

Also keep in mind the dangers of metal work. This is the most dangerous work you'll do to build the wind turbine. Metal is often sharp when you cut it. Grinding can be fairly dangerous—lots of shards flying around and we've seen grind stones explode before, so be sure to wear a face shield. Cutting with a torch or welding has its obvious hazards (molten hot steel and sparks flying). Just follow proper safety procedures and odds are you'll live through it! And be sure to review the metalworking section in Chapter 6, *Shop safety*, before proceeding. The completed frame is shown resting on a test stand at right in Figure 9.1.

Figure 9.1—Completed wind turbine frame. It's supported on a test stand.

Frame materials list

- 1/4 inch thick steel plate, 15 inches square
- 1 inch schedule 40 pipe, 9 inches long
- 2 inch schedule 40 pipe, 4 inches long
- 2-1/2 inch schedule 40 pipe, 12 inches long
- 3 inch schedule 40 pipe, 3-5/8 inches long
- 1/4 thick steel discs, 6 inches diameter, quantity 2 (not shown)

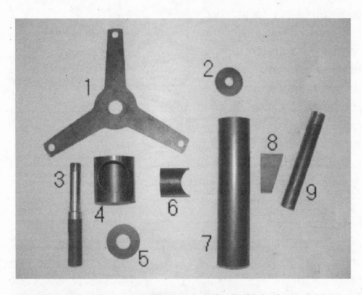

Figure 9.2—All of the cut metal parts needed to build the frame, laid out and numbered. Each piece has its own numbered section outlined in the text below.

Figure 9.2 at left shows all the parts you'll need to make (or have made) to build the frame. Not pictured are the parts of the tail itself which we will discuss later in the *Tail* chapter. Some of these parts are available pre-cut from a CNC CAD machine, check the *Sources* chapter for details on where to buy them. We build lots of these wind turbines, and have all of our parts cut this way. It saves many hours of cutting, drilling and grinding, but is fairly expensive.

1) Stator bracket

The stator bracket (part #1 in Figure 9.2, and with dimensions in Figure 9.3) is cut from 1/4 inch thick steel plate. It supports the front of the wheel spindle on which the main wheel hub/bearings mount, and it supports the stator. To lay it out, find center and draw two circles. The inner circle is 4 inches in diameter, the outer circle is 15 inches in diameter. Lay out the three "spokes" 120 degrees apart. The spokes are 1-7/16 inches wide at the outer diameter and 2 inches wide where they meet the inner 4 inch diameter circle.

Centered on each spoke at the 13-3/4 inch diameter is a 1/2 inch hole. These accept the studs that will mount the stator. In the center of the bracket is a 1-1/4 inch diameter hole that fits the wheel spindle. You can cut this out with a torch or a plasma cutter easily. If those tools are not available, you could simplify the design, keeping in mind that the critical measurements are the locations of the 1/2 inch and 1-1/4 inch holes. Some folks will make the inside 4 inch diameter circle with a hole saw, and then make the spokes with bar stock and weld them on. While it doesn't look quite as neat, it works just as well as

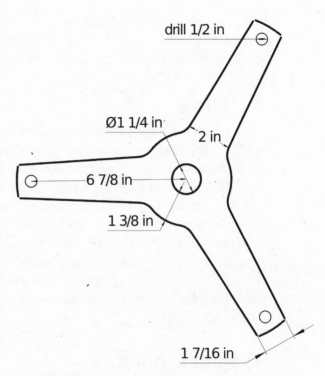

Figure 9.3—Stator bracket dimensions.

a fancy CNC-cut bracket. The spokes should be very rigid, so if you do try it with bar stock use at least 1-1/2 inch by 1/4 inch. Any thinner might bend under the torque that the stator will impose upon it.

2) Yaw bearing cap

The yaw bearing cap, part #2 in Figure 9.2, is simply a 2-3/4 inch diameter disc of 1/4 inch steel with a 3/4 inch diameter hole drilled in its center. It will be welded to the top of the yaw bearing, and the hole will accept the power cord from the alternator which will run down the center of the tower. If you purchased pre-cut magnet rotors, you'll already have a scrap that's 2-3/4 inches in diameter from the magnet rotor center hole and all you need to do is drill the hole in the center. If you are going to cut the magnet rotors yourself, your

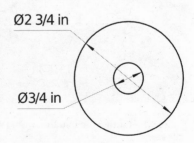

Figure 9.4—Yaw bearing cap dimensions.

scrap will still be usable if you do it with a hole saw. If you plan to cut your magnet rotors with a torch, the diameter of the scrap may be too small and you should fabricate this part separately.

3) Spindle

This is Dexter part #BT-8, and it's part #3 in Figure 9.2. This is the spindle onto which the wheel hub (Dexter part #81-9A) fits. It's probably the most common spindle in use for 1,000 pound trailer axles and it's widely available. There are other slightly cheaper spindle/hub setups available but we prefer this one. It's the best choice because it's common and is machined on both sides, which makes life much easier for us—everything lays flat against the hub right out of the box. With other brands you may have to machine one side of the hub flat, which requires a metal lathe. You can certainly modify the design to accept other brands of spindle/hub assemblies but that's one reason we think it's easier to stick with the Dexter parts here. Check out the *Sources* chapter for suppliers.

4) Spindle housing

This is the piece of 3 inch diameter schedule 40 pipe that houses all the parts of the spindle, shown as part #4 in Figure 9.2. It's 3-5/8 inches long, and we cut a 2-1/2 inch hole in it with a hole saw, shown in Figure 9.5. The hole can be cut on center, but we prefer to cut it slightly off center. Cutting it towards one side allows you to push the whole alternator forward slightly in relation to the yaw bearing, and gives a bit more clearance between the stator bracket and the yaw bearing. It also gives slightly more clearance between the blades and the tower. Use a high-quality bi-metal hole saw. Be sure to run the drill press at its lowest possible speed and use plenty of cutting oil.

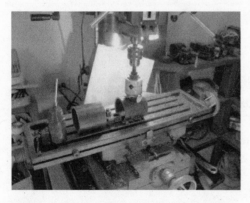

Figure 9.5—Cutting the spindle housing with a hole saw.

5) Rear spindle support

This disc is just under 3 inches in diameter (cut it out with a torch, or a 3 inch hole saw), and it has a 1-1/4 inch diameter hole in the center which fits around the back of the wheel spindle (part #3). It's shown as part #5 in Figure 9.2, and with dimensions in Figure 9.6. The stator bracket supports the spindle in the front, and this disc supports it in the rear.

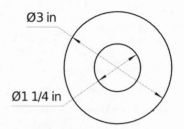

Figure 9.6—Rear spindle support dimensions.

6) Alternator offset bracket

This part connects the alternator assembly to the yaw bearing (part #7), and is shown as part #6 in Figure 9.2. It is cut from 2 inch diameter pipe. One end is coped with a hole saw so that it can be welded to the yaw bearing. The other end is left flat and fits into the 3 inch pipe spindle housing, which has a 2-1/2 inch hole cut to accept it (part #4 in Figure 9.2). When you cope the alternator offset bracket with a hole saw, the distance from the center of the hole saw to the other end of it should be 3-1/4 inches. Use a 2-1/2 inch diameter hole saw to cut this, and then touch it up a bit with a grinder so that it fits nicely against the yaw bearing for welding.

Figure 9.7—Alternator offset bracket, showing hole saw used for coping it.

7) Yaw bearing

This is the part that slips over the tower top, and is shown as part #7 in Figure 9.2. It's 12 inches long, and made from 2-1/2 inch schedule 40 pipe.

Holistic sawing

A quality hole saw is essential for boring into steel pipe. Thin, flexible saw sets with which you can interchange blade diameters into a flimsy hub will not do the job, nor will adjustable models. Look for a hole saw set that includes a sturdy arbor with retractable pins and solid bi-metal saw blades.

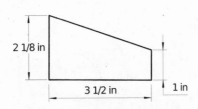

2 1/8 in

3 1/2 in

1 in

Figure 9.8—Tail bracket dimensions.

8) Tail bracket

The tail bracket sits between the yaw bearing and the tail pivot. It's also cut from 1/4 inch steel plate, is 3-1/2 inches tall, 1 inch wide at the bottom, and 2-1/8 wide at the top. It is shown as part #8 in Figure 9.2, and is dimensioned in Figure 9.8.

9) Tail pivot

Shown as part #9 in Figure 9.2. This part gets welded to the tail bracket. The tail slips over it, and "hangs" and pivots on this part. It is 9 inches long and made from 1 inch schedule 40 pipe.

Assembling the stator bracket and spindle

Figure 9.9—Parts to be welded for supporting the alternator.

Figure 9.10—Using a lathe chuck to hold the spindle square with stator bracket for welding.

Once all the pieces are cut out, you can begin welding the machine together. The first step is to build the assembly that supports the alternator. Once that's finished you weld the tail bracket to the tail pivot, and then you weld both of those to the yaw bearing. Start by gathering the parts pictured at left in Figure 9.9: the wheel spindle (part #3), the stator bracket (part #1), the 3 inch diameter pipe with the hole in its side (the spindle housing, part #4) and the 3 inch disk with the 1-1/4 inch hole in its center (the rear spindle support, part #5).

We like to use an old three-jaw lathe chuck as a clamp for this welding operation, but if that's unavailable a decent sized vise should work fine. You need to clamp the wheel spindle so that the machined part is facing down, and put the stator bracket around it as shown in Figure 9.10. Now 6-5/8 inches of the wheel spindle should be sticking up from the surface of the stator bracket. The stator bracket may not be perfectly flat, but try to get things as close to square as possible, so that the angle between the stator bracket and the spindle is always 90 degrees. First tack weld the spindle to the stator bracket in three or four small spots on different sides. This will hold it in position nicely so that you can really weld it on there solidly. If you don't tack weld it first and just start welding around the circle, it will be pulled out of square as you are welding, because welds shrink as they cool.

After you've tack welded the spindle to the stator bracket, go ahead and weld it there permanently, as shown at right in Figure 9.11.

Figure 9.11—Stator bracket welded in place on the spindle.

Now center the 3 inch pipe (spindle housing) on the stator bracket, shown at right in Figure 9.12. The height of the spindle should be the same as that of the pipe. The spindle housing has the 2-1/2 inch diameter hole you cut in one side. If you cut the hole off center like we described earlier (more towards one side then the other), then put the pipe on the stator bracket so that the hole is most distant from the stator bracket. If you cut the hole on center there's only one way to align it. Don't weld the spindle housing to the stator bracket yet. Where the 2-1/2 inch diameter hole points with regard to the spokes on the stator

Figure 9.12—Aligning the spindle housing on the stator bracket.

bracket is not terribly important, but we usually put it opposite one of the spokes. This way when the wind turbine is together, you'll have one of the spokes coming out from the stator bracket pointing exactly away from the yaw bearing. It just looks a bit neater and makes the machine less fragile if we have to ship it somewhere, but it doesn't affect the workability of the wind turbine.

Take the 3 inch diameter disc with the 1-1/4 inch hole in it (the rear spindle support), and position it inside the 3 inch pipe. It should fit around the back of the wheel spindle and inside the 3 inch pipe as shown in Figure 9.13. A big magnet serves nicely to hold it there

Figure 9.13—Using a magnet to clamp the rear spindle support.

Figure 9.14—Tack welds.

Figure 9.15—Final welds on the stator bracket and spindle.

Figure 9.16—Assembly propped up on shim.

while you tack weld it—but keep the magnet a bit away from where the actual weld will be made, or it will attract and splatter the molten metal.

Once everything looks aligned and centered, go ahead and tack weld it all together (3 inch pipe tack welded to stator bracket, 3 inch disc tack welded to 3 inch pipe and wheel spindle). The tack welds are again important to do first, or else the heating and cooling of the final weld will move things out of alignment. The tack welds are shown in Figure 9.14.

This is your last chance to make small adjustments in alignment, gently, with small raps from a hammer. After everything is lined up correctly weld it all together permanently, as shown in Figure 9.15. It's not necessary but will look much nicer if you take the time to grind your welds down and smooth things out. Now is a good time for this, as once you weld this whole assembly to the yaw bearing it will be difficult or impossible to access these welds with a grinder.

Next you'll weld the assembly that you just finished to the yaw bearing. There are a couple of weird angles involved here, and the distance between the yaw bearing and the wheel spindle is fairly critical. We've built a jig to make assembly of a wind turbine frame quite easy, and if you plan on building more than one machine such a jig may pay off. To do it without a jig, start by putting the spindle/stator bracket assembly that you just finished on your workbench, on top of a 1/2 inch thick shim (a piece of wood or 1/2 inch thick steel or whatever you have available) to prop up the spindle housing as shown in Figure 9.16.

The yaw bearing is 12 inches long. Mark the center of it at 6 inches. The layout is shown in Figure 9.17, and you are looking at it from the bottom of the wind turbine. You can see how the 1/2 inch shim under the spindle/stator bracket assembly pushes the alternator forward a bit. If you can imagine a vertical line through the center of the yaw bearing (the 12 inch by 2-1/2 inch pipe) parallel to the spindle, there should be 5 inches between it and the center of the spindle. In other words, when you are finished the center of the alternator will be 5 inches to one side from the center of the tower.

Figure 9.17—Laying out the alternator offset bracket and yaw bearing.

Figure 9.19—Welded together.

Now add another shim, this one 3/4 inch thick and under the *top* of the yaw bearing. Figure 9.18 shows the side view of this. This offset will tip the whole alternator back about 5 degrees. The wind turbine blades are then also tipped back and have a nice safe clearance between their tips and the tower.

Figure 9.18—Alternator offset bracket and yaw bearing layout, side view.

Tack weld this all together. Inspect it to make sure the offset between the spindle and the yaw bearing is correct, and that all other angles are correct. If all looks good then permanently weld it together. It should look like Figure 9.19 above.

Next you'll need the tail pivot bracket and the tail pivot (parts #8 and #9 as shown in Figure 9.2). Mark the center of the tail pivot (4-1/2 inches) and position the bracket so that the top (the 2-1/8 inch wide part) is at the half way mark on the tail pivot.

Figure 9.20—Tail pivot bracket and tail pivot.

Figure 9.21—Aligning the turbine frame at 45 degrees in a vise, side view.

Tack weld it at the top and bottom and then quickly on each side, as shown in Figure 9.20. If everything looks good, then weld it permanently. This needs to be a good weld with good penetration because the entire tail hangs on this part!

Figure 9.22—Aligned in vise, bottom view.

Take the main part of the wind turbine (now it's the "yaw bearing stator bracket spindle assembly") and clamp it in a vise so that it's tipped at 45 degrees, as shown from the side in Figure 9.21 at left and from underneath in Figure 9.22 above. Place the tail bracket and pivot assembly on top so that it sits pointing straight up. The entire frame is chucked at such strange angles so that this assembly will simply rest there for welding—it's at the very top of the photo in Figure 9.22. Tack weld it on there. Again, inspect it and make sure everything is straight and square. If it's all good then weld it there permanently.

Figure 9.23—Ready to weld on the yaw bearing cap.

This is probably the most critical weld on the machine! The tail can slam around on the tail pivot during high winds and this weld takes most of the abuse. This needs to be a high quality weld with good penetration or else you risk having the tail fall off, which can be disastrous if it hits the blades on its way off and down. We've never had this happen, but we always worry about it! If you make a good weld here, there should be no problem.

The second to last step in fabricating the frame is to weld the yaw bearing cap to the top of the yaw bearing, as shown in Figure 9.23 at left. At this point, it's very handy to fabricate a simple stand to support the frame. You'll use it while fabricating and fitting the tail, and while assembling the alternator. It simply requires a sturdy base and a stub of pipe to slip the yaw bearing over.

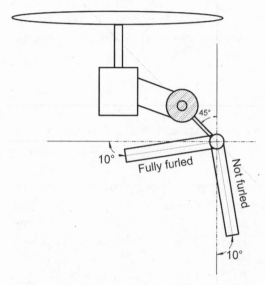

Figure 9.24 (above)—Completed frame, viewed from above. This is what your final product from this chapter should look like. Compare it to the drawing, too.

Figure 9.25 (left)—Completed frame, viewed from above, drawing. The tail is shown here also.

Figure 9.26—Completed frame, side view.

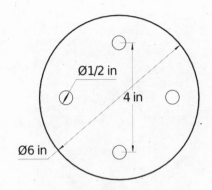

Figure 9.27—Front rotor hub.

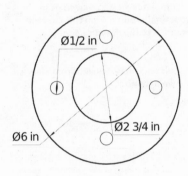

Figure 9.28—Back rotor hub.

From the top, your machine should now look like Figure 9.24 on the previous page. The drawing in Figure 9.25 shows it again from above, with the tail alignment angles shown. Viewed from the side, it should look like Figure 9.26.

Rotor hubs

Wait! Since you are still working metal, this is a good time to fabricate the rotor hubs that will reinforce the blades when you assemble them. You'll also use one of these parts as a jig when assembling the alternator. The front rotor hub is shown in Figure 9.27, the back one in Figure 9.28. The holes are on a 4 inch diameter. You can also purchase these parts pre-fabricated, check the *Sources* chapter.

For the final step you can at last shut down your welder, remove your hot and cumbersome welding mask, and crack open a relaxing beverage of your choice—see the *Glossary*, under "Fat Tire." Tap the hole in the middle of the yaw bearing cap you just welded on out to 1/2 inch – 14 tpi NPT (pipe thread) to fit a standard junction box wire clamp.

WELDER'S HAIKU

Joints, joints, joints, joints, joints

corner, edge, T-joint, lap and butt

After work, beer joint

Haiku by Jonathan Shipley

10. Tail

Now that you've fabricated the frame, it's time to build the tail vane. The actual mechanics of how the tail furls and comes back into position are confusing to many people, so we've tried to include as many pictures as possible of how the various parts align. When the tail is in normal operating position, it is tilted slightly up in the rear. This is simply for aesthetics and doesn't affect operation—we've found that tails that ride horizontally or slightly rear-down look "sad" and "droopy" when flying! In normal operating position, the tail is also angled ten degrees horizontally out from what would be straight on to the wind. The position of the notch that you will cut in the tail bearing controls this angle when you are building the tail. This 10 degree offset angle *does* affect operation; it is one of the factors that determines how straight into the wind that the turbine tracks during normal operation in slower winds. However, exact precision is not required (like with many aspects of this design). Anything from 5 to 15 degrees of offset will work fine, and would change tracking very little.

Figure 10.1—The tail boom is an easy metalworking project, and the tail vane allows you to express some creativity…in this case, perhaps a bit *too* much creativity!

When the tail is fully furled in high winds, it has swung both in and up from its normal position (Figure 10.2). The other side of the tail bearing notch controls how far the tail can swing in, and the tail stop assures that no matter what happens the tail cannot swing so far in that it hits the spinning blades from behind. The wind speed at which the tail starts to furl is adjusted by the weight of the entire tail boom and vane, the length of the boom and the surface area of the vane. The dimensions we give here are optimized for this particular 10-foot machine to make it begin to furl at just above 20 mph wind speeds.

Figure 10.2—A completely furled tail during high winds.

Tail materials list

- 1-1/4 inch schedule 40 steel pipe, 9 inches long
- 1-1/4 by 1/4 inch thick steel bar stock, 66 inches long
- 1 inch schedule 40 steel pipe, 60 inches long
- 1/4 inch thick Baltic birch plywood, at least 40 inches by 24 inches
- 5/16 inch – 18 tpi bolts, 1-1/2 inches long, quantity 4
- 5/16 inch – 18 tpi nuts, quantity 4
- 5/16 inch washers, quantity 8

Fabricating the tail boom

Figure 10.3—Laying out the tail bearing for notching. Two large magnets (right side) are used as clamps to hold it in place.

The tail bearing hangs and swings on the tail pivot, which is welded to the frame of the wind turbine. The pivot was attached to the yaw bearing via an angled bracket when you fabricated the frame. The bearing is made from 1-1/4 inch schedule 40 pipe. It's the same length as the pivot (9 inches). It slides over the pivot, and you will be notching it to fit over the bracket that attaches the pivot to the yaw bearing. One side of the notch will serve as a stop to hold the tail in its proper position. Figure 10.3 at left shows the layout for preparing to notch the tail bearing with a cutting torch. Two of your 1 inch by 2 inch magnets (used for building the alternator) stuck to the back of a piece of angle iron make for a nice jig/template to hold things in place and do a neat job.

Cut the notch halfway up the pipe (so that 4-1/2 inches is notched, and 4-1/2 inches is left undisturbed). In Figure 10.4 at right, George is cutting one side of the notch using the angle iron to help guide the torch. Once that side is cut, turn it about 160 degrees and cut the other side of the notch. We say to notch out "about" 160 degrees—this is the maximum width that you should make

Figure 10.4—Cutting the notch with a torch.

the notch. In most cases you could make it narrower (120 to 140 degrees is usually fine); this is not critical. If the notch is too narrow then the tail may not be able to fully furl. This depends on the width of the notch and also the thickness of the welds which attach the tail pivot to the tail bracket. When you reinforce the notch later in this chapter, you'll have the opportunity to make small corrections to the final size of the opening. Once both sides of the notch are cut, you can then cut between the lines and finish the notch. Save the scrap, you'll need it later.

Figure 10.5—Completed tail bearing notch, viewed from the notched end.

The tail bearing should look like Figure 10.5 (above right) after notching, looking up at it from the bottom (the notched end). There are are plenty of other ways to make this cut. If a cutting torch is not available, this notch is easy to make with a jig saw, hack saw, band saw, etc. A milling machine does nicely too, but it leaves a smaller scrap (the piece that comes out of the notch), and you do need that scrap later. We prefer to have the scrap larger and save time, so a band saw or torch is preferable. The notched tail bearing and the scrap you cut out are shown in Figure 10.6, at right. After cutting, use a hammer and beat the scrap a bit flatter so that its inner diameter fits nicely around the outer diameter of the tail bearing.

Figure 10.6—Notched tail bearing and scrap piece (save the scrap!).

Put the tail bearing on the machine and tack weld a 1 inch diameter plug (it could be up to 1-1/4 inches in diameter) in the top of the tail bearing. You should have this piece left over from fab-

Figure 10.7—Plug tack welded into place.

ricating the frame, if not you'll need to cut out another one. It needs to fit down inside the tail bearing so that the top of the plug is flush with the top of the pipe, and is shown in Figure 10.7 at right.

Figure 10.8—Tail bearing ready for tack welding the scrap piece on to it.

Turn the tail bearing all the way counter clockwise so that it comes to a stop against the notch on the side opposite, shown in Figure 10.8 at left. One side of the notch needs to come up against the tail bracket; this is the normal operating position.

Figure 10.9—Scrap piece welded to tail bearing to reinforce the notch.

Tack weld the scrap piece from the tail bearing notch to the tail bearing on the side opposite the alternator, as shown in Figure 10.9 at left. This serves as a reinforcement if the tail comes slamming down from the furled position back into the normal position. It will help keep the notched tail bearing from getting bent, or cracking around the notch. If you need to make a slight adjustment in the width of your notch, this is where you will do it.

The tail boom is fabricated from 1 inch schedule 40 pipe. It's 5 feet long. You need to cope one end so that it fits against the tail bearing. The angle between the tail bearing and the tail boom should be about 20 degrees. Figure 10.10 at left shows us cutting a 20 degree angle through half the pipe, leaving the other half cut at 90 degrees—a quick and cheap way to cope the pipe so it fits the tail bearing.

Figure 10.10—Cutting the 20 degree angle in the tail boom.

Figure 10.11 at right shows the finished angle cuts on the tail boom. The boom will now fit nicely against the tail bearing for welding.

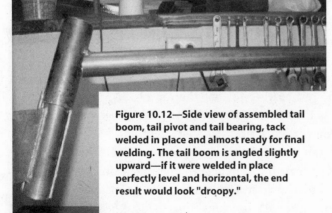

Figure 10.11—Completed angle cut on the tail boom.

Figure 10.12 shows how it should look from the side. There's about a 110 degree angle between the tail pivot and the tail boom. A common mistake can happen when folks weld the tail boom at 90 degrees to the tail pivot. The end result will be a tail that points slightly down and looks droopy and funny—it will work this way, but is unattractive. We prefer to have the tail pointing proudly upwards by a couple of degrees.

Figure 10.12—Side view of assembled tail boom, tail pivot and tail bearing, tack welded in place and almost ready for final welding. The tail boom is angled slightly upward—if it were welded in place perfectly level and horizontal, the end result would look "droopy."

Figure 10.13, next page, shows how the tail boom should be aligned, viewed from the top. When the tail is in its position of rest (the normal running position, not furled and turned all the way counterclockwise as far as it will go) it should be about 10 degrees from parallel with the wind (and the spindle). When fully furled (for protection in high winds, turned all the way clockwise as far as it will go), it should be about 10 degrees from perpendicular with the wind and the spindle.

Get someone to hold the tail so that it's in the non-furled position from Figure 10.13 and tack weld it there. Doing only tack welding at this stage allows you to correct the tail boom angle by bending things around a bit if necessary before the final welding.

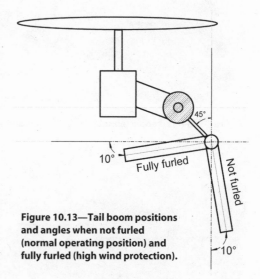

Figure 10.13—Tail boom positions and angles when not furled (normal operating position) and fully furled (high wind protection).

The reason the tail is sticking out at an angle about 10 degrees out of parallel with the wind direction is to compensate for the fact that the alternator is offset out 5 inches on the opposite side. When the wind is pushing against the blades and alternator, there is a force trying to turn the whole machine counterclockwise (assuming you're looking down at the machine). The tail is angled slightly to the other side to help counteract this force. That said, it's important to keep in mind that wind turbine blades *want* to run square with the wind, and tend to keep themselves pointed in that direction when running at proper speed. In our experience the 10 degree offset angle of the tail is not critical—anything between 5 and 15 degrees should work fine. Like most aspects of this design, it's not worth fretting a great deal over getting things perfect! If you are a couple degrees off it won't matter.

Now you can finish all the welds with a solid bead while the tack welds hold it all in place (Figure 10.14 at right).

Next, make a gusset 30 inches long out of 1-1/4 inch by 1/4 inch thick bar stock.

Figure 10.14—Final welds completed on the tail bearing and tail boom.

Figure 10.15—Tail boom gusset laid out for marking and cutting.

Figure 10.16—Gusset welded to tail boom.

Cut it to length (30 inches) and then lay it under the tail so you can mark with a soapstone where it needs to be cut (Figure 10.15 at left). Then cut it with a torch or a band saw so it fits well. This gusset probably isn't needed if the tail is welded to the tail bearing securely, but it looks nice and adds some strength—call it "insurance!" The 30 inch length is not at all critical.

Now you can weld the gusset onto the tail boom. It should look like Figure 10.16 at left.

Figure 10.17 below shows the "stop" (or "bumper") that you need to fabricate and weld to the tail boom. This will bump against the yaw bearing when the machine is fully furled to stop the tail at about a 10 degree angle from the blades (Figure 10.13, previous page) and thus ensure that the tail can never hit the blades from behind. The bumper is fabricated from 1/4 inch thick flat steel. The only critical measurement is the one that stands between the tail boom and the yaw bearing—it should be 2 inches if all the angles are perfect in your machine, but slight

Figure 10.17—Tail stop.

variations might mean you should change this. The stop shown is 2 inches wide at the bottom, 2 inches tall, and 1-1/2 inches across the top. It's angled on one side only for the sake of appearance.

In Figure 10.18 at right, George is clamping the bumper piece in there between the tail boom and the yaw bearing by folding the tail boom into fully furled position. It's helpful to have someone do this so you can position the part in its proper place for welding. Double check that the bumper piece is long enough to keep the tail out of the blades. It should be square with the yaw bearing and fairly well centered on it. The 1-1/2 inch long surface should touch the yaw bearing on center when the machine is furled. Once you get it aligned and clamped, tack weld it, check your work, and make the final weld.

Figure 10.18—George holding the tail boom in position for welding.

Figure 10.19, below left, shows the tail sitting in its normal running (unfurled) position. Figure 10.20, below right, shows the tail being held in fully furled position. Compare these photos to the drawing in Figure 10.13 (previous page). Note co-author Dan Bartmann's extremely grubby hand. This is a common hazard of metalworking!

Figure 10.19—Top view, tail sitting unfurled in normal operating position.

Figure 10.20—Top view, tail being held in fully furled position.

In Figure 10.20 you can see how the stop comes up against the yaw bearing. Notice the angle between the tail boom and the stator bracket—this same 10 degree angle will exist between the tail boom and the blades (Figure 10.13). Set up this way, the tail can never come all the way around and into the blades. In both Figure 10.19 and Figure 10.20, the "wind" is coming from the right side of the photo. You can see in Figure 10.20 how greatly the swept area is reduced by furling the blades at an angle to the oncoming wind.

Figure 10.21—George using a square to align the tail bracket.

The last metal part you'll need to fabricate is the tail vane bracket. It is 34 inches long and made from 1-1/4 inch by 1/4 inch thick bar stock. Each end has a 3/8 inch diameter hole drilled 1 inch from the end to hold the wooden tail vane. Use a square as demonstrated by George in Figure 10.21 at left. Clamp the tail vane bracket square with the tail boom and 9 inches from the end of the tail boom. Again—none of these measurements are critical, it's just our standard way of doing things. The exact measurements might change if you get creative with the shape of your tail vane.

If the machine is plumb on the stand, you can use a small torpedo level to make sure the bracket is vertical (Figure 10.22 at right). Once everything is aligned and clamped, tack weld it and inspect everything. Since it's only tack welded you can still make changes if needed. Once it looks good, weld it permanently. Now all the metal work is done, except perhaps for a bit of grinding to clean up the welds and sharp edges.

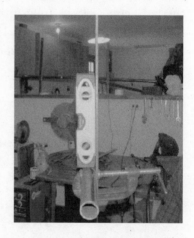

Figure 10.22—Using a torpedo level to ensure that the tail vane bracket is vertical for welding.

Fabricating the tail vane

Tail vane shape is not critical, so if you decide to make a different shaped tail for the machine it may make sense to change where the holes are drilled in the tail vane bracket for the best strength. Tail vane weight and surface area *are* critical, so for building this wind turbine don't deviate too far from the dimensions described here. To keep the machine furling early like it should, we like to keep the tail vane light in weight. We prefer 1/4 inch Baltic birch plywood for the vane, because it's very strong, light in weight and holds up to the weather very well if properly maintained. Plain old 3/8 inch exterior grade plywood can be used in a pinch, but it's heavier and will change the furling point toward higher winds, which is not desirable.

The tail vane should cover at least 5 percent of the swept area of the blades for this design. We usually make the vane 40 inches tall, 24 inches wide at the widest point, and slightly "arrow" shaped (because that looks cool, Figure 10.23). A tail shape that's more vertical than horizontal (like what we are building) seems to perform better with this particular wind turbine design. If you want to design your own creative tail vane shape, make sure it's at least 6 square feet in area, and be careful not to get too intricate—small, detailed areas may break off in high winds, and won't be visible from the ground anyway. Our friends and neighbors flying these wind turbines have built their own fanciful tails in the shape of whales, dogs, chickens, pigs, and worse (Figures 10.1, 10.24, 10.26 and 10.27).

Figure 10.23—A most attractive 10 foot wind turbine catching its first breeze! Note the sexy, arrow-shaped tail vane.

Figure 10.25—Rich mounts the tail vane of his new turbine onto the tail vane bracket.

Mount the vane securely to the tail boom with at least four 5/16 inch bolts, and be sure to use washers on both sides. Rich demonstrates in Figure 10.25. With the turbine frame on the assembly stand, slide the tail into place and observe how it will furl in the wind! This should clear up any doubts you or your mystified neighbors had about how a furling tail functions.

Figure 10.24—George's pig-shaped tail vane. He even painted it pink!

Figures 10.26 and 10.27—Matt's big wind turbine, and his border collie Dixie posing by the tail vane.

11. Magnet rotors

The wind turbine design in this book uses two magnet rotors that are directly attached to and spin along with the blades. They are made from 1/4 inch thick steel discs, with 12 large neodymium-iron-boron (NdFeB) "rare earth" magnets spaced evenly around each disc. Each magnet faces its opposite on the other magnet rotor—the north magnetic poles of every other magnet on one rotor are aligned with (and ferociously attract) all the alternating south faces on the other rotor, and vice versa. The stator, which you'll be building in the next chapter, is "sandwiched" between the two spinning magnet rotors, and doesn't move. The distance between the magnet faces on each rotor is called the "air gap" of the alternator. The back magnet rotor is shown in Figure 11.1 and the front magnet rotor in Figure 11.2. Figure 11.3 shows what the alternator would look like from the side. The front rotor has four extra tapped holes in it for "jacking screws," which allow you to gently lower it down into position against the tremendous magnetic force between the two rotors.

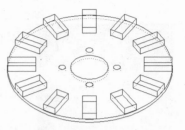

Figure 11.1—Completed back magnet rotor.

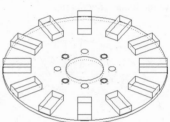

Figure 11.2—Completed front magnet rotor.

Once the alternator is assembled, the magnet rotors pull only at each other. But as you're building the magnet rotors, they'll try to grab any ferrous object in the room! Both the magnet rotors are extremely dangerous items. They'll pull steel tools and other magnets right out of your hand from a surprising distance before you even realize what's happening. If the two magnet rotors ever get

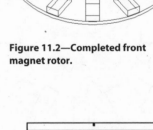

Figure 11.3—Side view of alternator.

anywhere near each other, they'll slam together so hard and fast that you may never get them apart—and if your hand happened to be in there, you'd lose fingers. Really! The magnet rotors are *that* powerful! Before you even start this phase of building a wind turbine, prepare a safe magnet rotor storage facility with no ferrous objects nearby, and where children, pets and casual passers-by won't come anywhere near them. Hang them on nails in the wall, on opposite sides of the room.

Building the magnet layout template

Figure 11.4—Aluminum magnet layout template.

You'll first need to make a new tool for this step—a template for placing the magnets. It's almost impossible to place the magnets in perfect alignment without one. The template pictured in Figure 11.4 was made at a local machine shop with a CNC water jet cutter out of 1/8 inch thick aluminum, and the cost was quite reasonable. See our *Sources* chapter for information on where to purchase a template if you don't want to make one. You can also build it from thin plywood or plastic, which are easy to cut with a jig saw or coping saw, but we like the aluminum because it won't break and is infinitely re-usable.

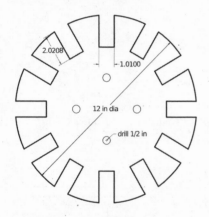

Figure 11.5—Magnet layout template dimensions.

The dimensions of the template are shown in Figure 11.5. It's a 12 inch diameter disc, with four 1/2 inch holes on a 4 inch diameter (just like the magnet rotors), and 12 equally spaced cutouts just slightly bigger than the size of each magnet (1 inch×2 inch×1/2 inch in this case). Be sure to read the *More powerful alternator sidebar* in Chapter 12, *Stator*, before buying or diving into building your magnet rotors! There are some possible design modifications that you might want to consider, including using round magnets instead of rectangular.

Magnet rotors materials list

- 12 inch diameter steel disc, 1/4 inch thick, quantity 2
- 1 inch×2 inch×1/2 inch N40 grade NdFeB magnets, quantity 24
- Cyanoacrylate glue (Superglue®) and accelerator
- Fiberglass cloth or mat, 2 square feet
- Casting resin, 1/2 gallon (see the *Resins* sidebar, later in this chapter)
- Resin filler (see *Resins* sidebar)

Fabricating the steel rotor discs

Start with your two steel discs, 12 inches in diameter. When you complete this step, each disc will end up with four 1/2 inch holes on a 4 inch diameter circle (a touch larger on the four holes to provide some clearance is nice) and a 2-3/4 inch hole in the center. These discs are also available already fabricated for purchase, see the Sources chapter. If you will be fabricating them yourself, you can build a rotating jig and do a pretty neat job cutting these out with an oxy-acetylene cutting torch, but we have ours cut out by a fabrication shop. A CNC laser cutter, plasma cutter, or water jet cutter will do a very nice job. If you have it done for you, you might also have the shop cut all the holes for you—it saves a lot of time and assures that things are accurate and on-center.

The dimensions for the front disc are shown in Figure 11.6, and the back magnet rotor disc dimensions are shown in Figure 11.7. If you decide to machine your own rotors, the first step after making the discs is to cut the 2-3/4 inch hole in the center of both rotors. Use a high quality bimetal hole saw and a drill press for this. The drill press needs to be run very slowly and you need to use lots of cutting oil. It is easiest to clamp the discs together and cut both at the same time, or lightly tack weld them together to be later separated using a cold chisel or grinder. Save the scraps from the center, they can be used in building the frame. Next, keeping both discs clamped together, drill out the four 1/2 inch holes equally spaced around a 4 inch diameter circle. The easiest way to do this is to put the wheel hub on the rotors and clamp it there, and drill right through the holes on the wheel hub. This saves a lot of layout and assures accuracy.

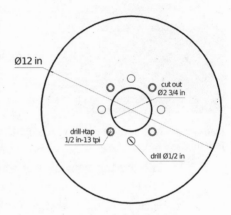

Figure 11.6—Front magnet rotor dimensions.

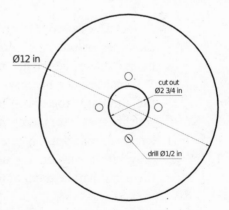

Figure 11.7—Back magnet rotor dimensions.

One of the discs needs four more holes drilled, 7/16 inch diameter (actually 27/64 inch, since it will be tapped) and also spaced around the same 4 inch diameter, located halfway between the 1/2 inch holes you just finished. You'll then tap these for the 1/2 inch–13 tpi jacking screws, which are used to assemble and disassemble the alternator. Again, we usually have these holes made at a fabrication shop—when they cut out the discs for us it's easy for them to use the same CNC machine to make all the holes, and the resulting product is perfectly aligned.

Figure 11.8—George tapping the jacking screw holes.

In Figure 11.8 at left, George is tapping the four smaller holes to 1/2 inch–13 tpi. It's important to use lots of cutting oil (or thread tapping fluid) when running the tap in. Try to keep the tap as straight as possible. Once it starts to cut threads, turn it just until things start getting tight, then back up a bit and "break the chip." Continue this until the tap goes all the way in and spins freely. Never force the tap in if things get too tight—always back it up, break the chip and then go forward again.

Use a countersink to chamfer the edges of only the 1/2 inch diameter holes, shown at right in Figure 11.9.

Figure 11.9—Chamfering the holes with a countersink.

This makes assembly easy, and helps protect the threads on the studs that hold the alternator together. Once this is done, all the metal work is finished for your magnet rotors. Both rotors will be greasy from fingerprints and cutting oil left over from drilling and tapping, so you'll need to clean the steel discs carefully with a solvent. We use lacquer thinner. After that, try to keep grease off them and handle them only with clean hands.

You're about to start putting magnets on the discs, so this is a good time to clean and tidy up your entire work area. Metal chips from the drill press and grinder will stick to the magnets and rotors during magnet rotor assembly, so either clean up thoroughly or move to another, cleaner work area for the next step.

The steel discs don't always arrive perfectly flat. Larger sheets of steel get bent in shipping and handling by forklifts and such, and even after the discs are cut out they may be slightly warped. Check for this with a straight edge and flashlight, as shown in Figure 11.10 at right. It's possible to flatten the discs, but it can be tricky. Instead we simply locate the dimension in which the disc is warped, and mount our magnets to the most

Figure 11.10—Checking for disc flatness using a flashlight and straightedge.

convex surface. The surface facing up towards the straight edge in the picture is the surface we'd put the magnets on. The tiny amount of warp will not affect the final product.

Now put the magnet template down on one of the steel discs and line the 1/2 inch holes up (Figure 11.11). Place the other steel disc down on top of that, with the holes lined up. Pin the "sandwich" together with two 1/2 inch drill bits (or wooden dowels, or bolts, or whatever, Figure 11.12). Looking

Figure 11.11—Template on first magnet rotor disc.

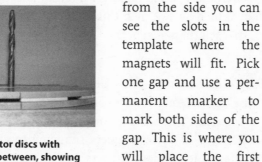

from the side you can see the slots in the template where the magnets will fit. Pick one gap and use a permanent marker to mark both sides of the gap. This is where you will place the first magnet on each disc. You can see the slots for the magnets and the marker lines in Figure 11.13.

Figure 11.13—Magnet rotor discs with template sandwiched in between, showing alignment marks.

Figure 11.12—Second rotor disc in place on the alignment pins.

Because the two magnet rotors must be perfectly aligned for final assembly, you should carefully index the rotors together. Use a small drill bit (3/16 inch is a good size) and drill a shallow divot (a dent, not a hole, don't drill through) into both discs between the marks you just made. These divots will be on the outside of the magnet rotors and will serve as index marks, so

Figure 11.14—Making the top index mark.

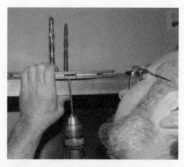

Figure 11.15—Making the bottom index mark.

that when you assemble the machine you will know how to line up the two rotors. Markers or paint don't do the trick—they tend to disappear as the piece is worked on, or get forgotten and covered by the final paint job. George is demonstrating how to make both index mark divots in the photos here, the top disc at left in Figure 11.14, and the bottom disc at right in Figure 11.15.

Once you've done all this, you can take the top rotor back off the stack and put it aside in your safe storage place—away from the bottom rotor because you are about to start playing with magnets!

Placing the magnets

For this alternator you'll need 24 NdFeB (neodymium-iron-boron) magnets, 1 inch by 2 inch by 1/2 inch thick. Check our *Sources* chapter in this book for retailers that sell these magnets.

CAUTION! *These are very powerful magnets and need to be treated with extreme care!* Please re-read our section on magnet safety in the *Shop safety* chapter before proceeding. Be aware of any serious health hazards involved, such as pacemakers. Two of these magnets snapping together on your finger could be very painful and leave blood blisters. We've been there and done that! Once these rotors are assembled, the possible attraction force and potential for injury are multiplied greatly, to where bones could be crushed and injury be permanent if the two magnet rotors came together with any of your body parts in between. Always be aware of the danger of assembled magnet rotors. This is why you cannot buy pre-assembled magnet rotors—they would be difficult to ship safely, and dangerous to unpack!

Build only one magnet rotor at a time. When it's finished, put it in a safe place. Hanging high on the wall, out of casual reach, is good. When building these magnet rotors, be sure that all ferrous materials (anything containing iron, which includes steel tools, wrenches, knives, scissors, razor blades, car keys, etc.) are away from the work area. Handle only one magnet at a time, and always grip it firmly. If a magnet is pulled out of your hand onto a piece of steel or into another magnet at high velocity, it may break and send shards flying! Store the magnets in a safe place, away from kids and folks who don't realize what they might be getting into. Keep them away from electronics, video tapes and other forms of magnetic storage medium. These large magnets are perfectly safe when handled properly, but most folks are not familiar with the dangers and there can be surprises—like when the magnet is yanked out of your hand to another magnet 8 inches away!

Please *again* re-read our full magnet safety warnings in the *Shop safety* chapter. All of this re-iteration of our magnet safety warnings may be starting to sound silly to you by now, but we speak from long experience in the wholesale, retail, and industrial magnet supply business! Magnets will bite you when you least expect it, and assembled magnet rotors will crush you.

Handling large magnets

The first thing you'll notice is that your magnet blocks arrive packed in a box with a steel liner, to comply with shipping regulations. Each magnet is separated from its neighbors by a plastic or wooden block. This is crucial! Without the separators, the magnets are extremely difficult to remove from each other. These magnets are so strong that they can be tricky to separate off of the stack, despite the spacers.

The best way is to place the stack on a wooden workbench and hold the stack firmly with one hand. Grasp one magnet firmly with the other hand and slide it off. You will not be able to just pull them apart, you have to "shear" them apart sideways. The procedure is demonstrated by expert magnet wrangler Rich in Figure 11.16. If you accidentally get two of these magnets stuck together without the spacers in between, you are in for a long, frustrating session in getting them apart again—the procedure shown in Figure

Figure 11.16—Expert magnet wrangler Rich, separating magnets off the stack.

11.16 is still the correct way to do it, but it will be much more difficult. Remember that you must use a wooden workbench, not steel! After separating a magnet for insertion, keep it far away from the stack, the rotor, and any other ferrous metal. Always grip it firmly. Use zen-like concentration while it is in your hand. It's not a bad idea to order a couple extra magnets in case you break one, and for general use around the shop.

Figure 11.17—Placing the first magnet on the rotor.

Now you can place the first magnet on the bottom magnet rotor. The template is pinned to it and made of wood or aluminum so it won't be attracted by the magnets. But, each magnet is strongly attracted to the steel disc, so you need to hold the disc down with one hand. While firmly gripping the magnet in the other hand, bring it towards the edge of the rotor and slide it into the slot. Don't just try to set the magnet down into the template—it will be yanked out of your hand and hit the rotor hard, possibly breaking the magnet! In Figure 11.17, Rich is demonstrating the proper magnet placement technique of sliding it in there while the magnet's pull does the rest. It will pop right into place.

Figure 11.18—Rich testing for polarity before inserting the next magnet.

The magnets need to be spaced around the disc with alternating poles facing up. All magnets have two poles, a north and a south. Opposite poles attract one another, and like poles repel. It doesn't matter how you put the first magnet down so long as the poles alternate from there—north and south are irrelevant as long as the two rotors are aligned in opposite polarity. The safe way to place the rest of the magnets is as follows, and is illustrated in the photo at left, Figure 11.18.

- Hold the magnet rotor down firmly to the work bench with one hand, which should be placed over the magnet that's next to the one you are about to place.

- Holding the next magnet firmly, bring it over your other hand which is holding down the rotor. If the bottom of the magnet in your hand is repelling the one on the rotor next to it, then slide it into the slot carefully in its current orientation. This is because you know that if the bottom of the magnet in your hand is repelling the top of the one on the rotor, then you have like poles facing each other.

- Finally, perform the "hand test" again after all the magnets are placed. The poles should alternate as you move your hand around the circle.

Once all the magnets are placed on the first rotor and tested for polarity, you can remove the pins and pry the template off. Do this carefully so the magnets don't slide around. Run a bead of thin viscosity cyanoacrylate glue down both sides of each magnet, where they meet the steel (Figure 11.19). The glue will wick underneath the magnets and hold them in place until they are solidly cast in resin. Large bottles of cyanoacrylate (2 ounces usually) are available, see the *Sources* chapter for more information. Avoid supermarket glues—they are not as thin and don't work nearly as well. It's also handy to have accelerator which will force the glue to harden immediately. The accelerator usually comes in a small spray bottle. We don't rely on this glue to hold the magnets down forever, it's a temporary means to keep things in place until we finish the casting.

Figure 11.19—Tacking the magnets into place with cyanoacrylate glue and accelerator, after removing the template.

Take a roll of fiberglass drywall tape (this stuff is sticky on one side) and cut the roll with a razor knife so that you can peel off a strip of the tape about 1/2 inch wide, shown in Figure 11.20. Wrap the tape around the edge of the magnet rotor several times. Be sure that none of the tape sticks up above the top of the magnets. You might also check out the sidebar at the end of this chapter on banding the magnet rotors with stainless steel—that's an alternative method of strengthening them that we have recently started using.

Figure 11.20—Cutting drywall tape for bolstering the edges of the casting.

The assembled rotor should look like Figure 11.21. Now that the first magnet rotor is finished and ready for casting, hang it in your safe magnet rotor storage facility. If you must grind or weld nearby, cover the rotor with a trash bag.

The second magnet rotor is built just like the first, but you must take care that the index mark magnet is placed with the right polarity. Put the template down

Figure 11.21—Magnet rotor ready for casting!

so that the four holes line up, and one of the slots lines up with the marks you made earlier. This assures you that the magnets you place will be facing each other with correct polarity when the rotors are assembled. The top of the first magnet on this rotor must be the opposite pole as the top of the first magnet you placed on the first rotor. In other words, each magnet on each face must attract its opposite. Once you get the first magnet down, follow the same procedures above as you did to assemble the first magnet rotor.

Check your work!

You can easily double check your work now. This is the time to find errors in magnet placement, instead of after the resin has been poured! Take a small magnet and hold it in your hand (don't turn it over—hold it in the same position always for the testing). Each magnet rotor has one magnet (the first one you placed) between the index marks you made. The test magnet in your hand should attract this index magnet on one magnet rotor, and repel it on the other. Then you can once again go around each magnet rotor, and the test magnet should attract one magnet, repel the next one, attract the next one, and so on. If you made a mistake, you need to carefully remove the offending magnets, put the template back on, flip the misplaced magnet over, and get it right. Once all the magnets are tacked down with Superglue®, it's a good idea to clean the magnets and the rotors one more time with lacquer thinner to make sure there's no grease. Grease interferes with the casting resin adhering to the magnets and rotor disc.

Preparing for casting

Figure 11.22—Fiberglass cloth cut outs to strengthen the rotors.

Our design uses fiberglass cloth from the auto parts store to help strengthen the castings. Cut out two rings from fiberglass mat or fabric. They should be 12 inches in diameter, with a 6-1/2 inch diameter hole in the center as shown in the photo at left, Figure 11.22. Grease the interior of the mold everywhere with mold release compound. Car wax or Johnson's wood wax make good mold release compounds. We've also used shortening from the kitchen and axle grease (axle grease is kind of disgusting and messy but it works). Grease it really well, especially the first time you use the mold. The first coat tends to soak into the wood, but after several applications it gets better. The point here is to make the mold greasy so the resin won't stick to it. Then run a bead of caulk around the outside of the 12-1/2 inch diameter hole in the mold. Also run a thin bead around the outside of the 1/2 inch thick, 6-1/2 inch diameter disc "island." Insert the 1/4 inch drill bit into the center hole to use for alignment. When you are ready to cast, your mold or molds should be laid out as shown in Figure 11.23. Note the beads of caulk.

Figure 11.23—Molds, ready to go.

Drop one of the magnet rotors into the mold carefully. It should fit nicely on the smaller 2-3/4 inch diameter disc in the center of the mold so that the magnet rotor is centered, shown in Figure 11.24 at right.

Figure 11.24—Rotor in place in the mold.

Now put the 6-1/2 inch diameter "island" disc down. The drill bit will serve to center it on the disc. The side that you've run caulk around should face down, and you'll need to

Figure 11.25—Ready for casting, with "island" in place. Note the bead of caulk around the rim.

press it down onto the magnet rotor. The caulk will assure that no resin can run under it. Refer to the photo at left, Figure 11.25.

It's now time to pour the resin!

Casting

You will be using resin to cast the rotors. Refer to the *Resins* sidebar later in this chapter for information on what to buy. It's best to buy it by the gallon (it takes almost exactly one gallon to build the entire alternator). The resin will come with the proper hardener in small plastic tubes, and the hardener is nasty stuff. It smells bad, and the fumes are toxic. It's best to work outside or in a very well ventilated area. Be sure to wear safety glasses, rubber gloves, and a respirator. It takes almost exactly 1 quart of resin to make one magnet rotor.

Usually a gallon of the resin comes with two tubes of hardener, each containing 0.77oz (22ml). When casting resins, they tend to heat up and get hard much faster than they would in normal applications—especially if it is

Resin roundup

When building the magnet rotors and the stator, we frequenty speak of "resins" with which we cast parts into assemblies. The resin is a liquid polymer that comes with separate "catalyst," and when the two are mixed together the liquid goo will set up into a solid plastic after a certain amount of time. Since the magnet rotors will be your first castings, this is a good time to discuss different types of resins that could be used, their properties, and their advantages and disadvantages for this wind turbine application. Like many of the topics covered in this book, one could easily write an entire book on resins alone! The use of modern polymers is a science all to itself. The point here is to give some basic information and discuss our own experiences and choices a bit.

No matter what resin you use, a few things always hold true:

• Read the manufacturer's instructions carefully.

• Read and follow the manufacturer's safety guidelines *exactly*.

• All of these resins are very messy.

• All of them are toxic.

• All come in two parts, and once mixed together you have limited working time.

If things are too warm (the environment, the mold or the parts) or if you use too much catalyst (or sometimes even if the volume of the casting is too large), the resin can get quite warm or even toasty hot during the cure. This often results in cracking or warping. In some cases these resins can get so hot while curing that they ignite! It's very important to read the instructions and work in a safe place with good ventilation.

There are three main types of resins usable for your magnet rotor and stator castings: polyester, vinyl ester, and epoxy. You'll need to make your choices based on their properties, their cost, and perhaps their availability. If you have internet access it becomes easy to procure resins that may not be available locally. If you don't have access to the internet, there are probably products that will do the job available locally at auto parts or hardware stores. To build the 10 foot diameter wind turbine described in this book you'll need about 1 gallon of resin.

Polyester

Polyester is the most common resin available and it's the least expensive. Many folks call it "fiberglass resin" because it's used in fiberglass layup, but it doesn't have any fiberglass in it at all. Many other types of resin can be used in the same applications, so the term "fiberglass resin" doesn't really make much sense! You can usually find it in gallon containers at auto parts stores and hardware stores. The catalyst used is methyl ethyl ketone peroxide (MEKP for short). Generally the catalyst is mixed in a ratio of about 1 percent with the resin. The resin will set up faster with more catalyst, or slower with less. Polyester works pretty well for casting stators. It holds up to fairly high temperatures and it's inexpensive. It does tend to shrink quite a bit when it cures, and it's a poor adhesive. Both of these properties make it less suitable for casting magnet rotors. During curing we've often seen polyester resin separate from the magnets and the steel rotors. Given time with sun shining on it we've seen it crack and break off magnet rotors all together. We've never had repeated major problems using polyester to make a stator, but we've had a few failures on magnet rotors. While polyester will work OK in all of these applications it's never our first choice.

Advantages of polyester resin: Easy to find anywhere in the world, easy to use, inexpensive, holds up to fairly high temperatures.

Disadvantages of polyester resin: High shrinkage, only moderate strength, lousy adhesive properties, very toxic.

Vinyl ester

Vinyl ester is more difficult to find than polyester, and we usually have to mail order it from the internet. See our *Sources* chapter for information on where to purchase vinyl ester resin. It generally costs a bit more than polyester, but the price is similar. It's much tougher than polyester and it withstands higher temperatures. It shrinks far less in the mold than polyester. The catalyst is the same (MEKP) and it's very much the same to work with: It smells the same and it's just as toxic. It's a much better adhesive than polyester, and we've been using it for stators and magnet rotors for some time now with no problems yet.

Advantages of vinyl ester resin: Fairly low cost, easy to use, lower shrinkage than polyester, high strength, holds up to very high temperatures, better adhesive than polyester.

Disadvantages of vinyl ester resin: Harder to find than polyester, very toxic.

Epoxy

There are several different types of epoxies designed for different applications and different cure times. They can vary in cost greatly. Epoxy also always comes in two parts, a resin and a "hardener," and the ratio in which they are mixed is critical. Usually it's either a 50:50 or 75:25 mix of resin and hardener—this varies between types and manufacturers, so following instructions exactly is critical. The big advantage of epoxy is that it's an excellent adhesive, but there are different types and not all epoxies are suitable for the sort of casting you'll need to do for a wind turbine alternator. We've tried cheap supermarket and hardware store epoxies on magnet rotors, and although they adhered very well to the magnets and the steel, they cracked badly while curing. At this time, the only epoxy we feel very confident in is West System®, see our *Sources* chapter for more information on where to buy it.

West System® epoxy is actually our first choice for casting magnet rotors, but the high cost (about US$100 per gallon) is often prohibitive. Epoxy shrinks the least of all three resins when curing. It does not withstand high temperatures as well as either polyester or vinyl ester. For stators, epoxy would be our last choice, but for magnet rotors it's excellent.

Advantages of epoxy resin: Very low shrinkage, excellent adhesive, high strength, less toxic than other resins.

Disadvantages of epoxy resin: High cost, critical mixing ratios, not as resistant to high temperatures. Some people are *very* sensitive to the hardener, and can have severe reactions to it.

Fillers

It can be advantageous to add certain fillers to the resin—they can add strength, improve thermal conductivity and reduce cost. There are several choices that could be used with any of the resins listed above. Since fillers cost less by volume than resin, their use can save money as less resin is required when you add a filler. Below we'll discuss a couple of the fillers we've tried, and our thoughts regarding them.

Chopped fiberglass is available from most suppliers. It's available in different lengths, with the fibers ranging from about 1/8 inch to 1/2 inch long. It's simply little strands of fiberglass, but adds a great deal of strength to the casting. We've built many machines without it and would say it's optional—but a bit more strength never hurts!

Talc, or "talcum powder," is a very inexpensive filler that works well with all these resins. It can be ordered from resin suppliers, or just go to the supermarket and buy some baby powder! Baby powder is usually just talc with a funny smell mixed in, just like a baby's bottom. Mixed by volume of about 50:50 with the resin, it will save a good bit of resin and improve the thermal conductivity and strength of the casting. The use of talc as a filler also reduces the amount of heat in the casting while curing, which makes cracking and warping less likely.

The drawback of using talc (or any powdered filler) is that it makes the resin thicker and increases the odds of trapping air bubbles in the casting. While the casting cures, these air bubbles rise to the top and leave holes in the top of the casting. This is usually only a cosmetic defect. We've made many stators with and without any filler, and either way seems to work fine.

Alumina trihydrate (ATH) is an exellent and inexpensive filler that has worked very well for us. It's made for large castings, and it improves thermal conductivity to reduce heat buildup while curing in the mold. This in turn means that it also improves the thermal conductivity of the finished casting. In our experiments, ATH seems so far to be an ideal filler for use in stators, and we've been using it on our most recent castings. It's employed in the same way as talc, and is available from most businesses that supply resin.

Conclusion

For most of the 10-foot diameter wind turbines we build, both the magnet rotors and the stator are cast in vinyl ester resin, with ATH filler. For larger projects like our 17-foot and 20-foot models, we cast the magnet rotors with West System® epoxy, and the stator with vinyl ester and ATH. If you are in a remote part of the world (or on a tight budget), plain old Bondo® brand liquid polyester resin from the auto parts or hardware store will work just fine for you, with baby powder as a filler if you choose to use one.

warm outside and if the resin is warm to start with. We use about half the hardener that the instructions call for. This lets it harden more slowly—we believe it makes the casting stronger, it shrinks less, and the final casting will be less likely to crack. There have been times when we've used half the tube for 1 quart (exactly what the instructions call for) on warm days and the resin has become hard in 15 minutes or less—a big problem since it was hard before we could even finish pouring it. If it goes off too quickly, there is also the risk of it catching fire! So be careful with your mixing. If you'd like, there are powders available to color the resin, or you can just use a little bit of acrylic enamel paint to give the resin a color. We sometimes use about 1 part paint to 50 parts resin. This is entirely for cosmetic reasons.

Use disposable plastic containers for mixing the resin, hardener, and coloring compound. If you are using a filler (see the *Resins* sidebar), this is when you add it. Stir thoroughly with a disposable plastic rod or spoon. The plastic magnet separators your magnets were shipped with make excellent stirring sticks.

Level the mold on your table, and pour resin in over the tops of all the magnets, as shown at right in Figure 11.26. The mold should be completely filled with resin. Wear latex gloves! Wear eye protection! Wear a respirator if you are doing it inside!

Place the fiberglass ring over the top and work it in with a stick so it becomes saturated with resin (Figure 11.27, at left). Work the air bubbles out as best you can. Pour a bit more resin over the top and work that in (Figure 11.28, at right). At this point it doesn't hurt to beat on the mold or vibrate it (with a vibrating sander or something similar) to work air bubbles out. Air bubbles won't really hurt it, but they don't look nice. We always get a few.

Figure 11.26—Pouring resin into the mold.

The lid also has a 1/4 inch hole in the center. Place it down over the drill bit and on top of the magnet rotor. The best way to clamp it down is with C clamps, though bits of steel (wrenches, etc.) will also work because they'll all stick to the magnets in the ro-

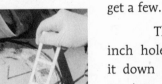

Figure 11.27—Working the resin into the fiberglass ring.

Figure 11.28—Pour more resin, get the air bubbles out.

tor, and the cyanoacrylate glue on the magnets *usually* keeps them in place. In Figure 11.29 at right, we used extra magnets as clamps. Keep an eye on the resin that spills out of the mold. When it starts setting up and getting rubbery, clean it off the outside of the mold with a utility knife—this way you don't have to chip off hard resin later. Don't take the lid off, though, until you feel the resin is good and hard. In practice, depending on the temperature and the amount of hardener you used, this may take anywhere from 1/2 hour (which is hot and fast enough to be scary—expect cracks and shrinkage when it goes off that fast) to 24 hours. The slower the better with regard to preventing shrinkage and cracking.

Figure 11.29—Mold lid clamped in place with magnets.

After the resin is completely set up you can remove the lid from the mold. If you made the mold well and greased it well, the rotor will just fall out of the mold when you turn it over. If it doesn't, gently tap it on the back with a hammer or rubber mallet and the casting should come loose—but be careful of the magnets attracting the hammer! Sometimes things get tricky and you'll have to pry it out, or even take the mold apart. The wooden disc on the inside of the rotor should knock out easily with a hammer through the hole on the back side of the magnet rotor.

The edges of the rotor casting will be rough when you get it out of the mold (Figure 11.30). You can cut the excess resin off with side cutters, or remove it with a sander. A belt sander or orbital sander works very well, but be sure to wear a dust mask and be aware that the magnets in the rotor will try to grab your sander right out of your hands! Clean up all the burrs so that nothing sticks up beyond the surface of the magnets.

Figure 11.30—Fresh out of the mold.

Now you have a finished magnet rotor (Figure 11.31)! Repeat the process with the second one in the same mold. We actually use two molds (molds are easy to make) so that we can get all this done in one shot, but if you're not in a rush one at a time works fine.

Hang your completed magnet rotors up in your safe storage facility, and cover them with plastic trash bags to keep metal chips away. Next, you'll be winding coils and casting the stator.

Figure 11.31—Completed magnet rotor.

Banding the rotors?

One alternative to casting the rotors as described in this chapter is to band them with stainless steel around the edges. We have recently started building all our turbines this way, and have been building larger machines this way for years. The process of building molds and casting the rotors as described in this chapter has the advantages that the magnets are totally encapsulated by resin, and it's relatively simple to do with simple tools. Banding the rotors has advantages too: no molds are required, not having cloth and resin on top of the magnets gives a bit more mechanical clearance between the rotors and the stator, if something really goes wrong it's very unlikely that the magnets could ever come out, and it looks nice.

Making the band

First, place the magnets on the rotor as described in this chapter. Obtain some 3/4 inch wide x 0.03 inch thick stainless steel strapping, it's available from most companies that provide shipping and packing materials. Cut the band so that it's 1/8 inch shorter than the circumference of the steel rotor. A jig or clamp of some kind is necessary to hold the ends of the band perfectly straight with one another, and ours is shown in Figure 11.32. The bottom of the jig (underneath where you'll be welding) should be copper so that the weld will not stick to the jig, and the copper helps remove heat from the stainless during welding to reduce the chances of blowing holes

Figure 11.32—Jig for welding the stainless steel band ends together.

in the material with the welder. We normally use a wire feed welder with stainless steel wire for this. A TIG welder would probably work better, and we have also seen folks braze or silver solder the band together. Once the band is welded, grind down the weld and inspect it for strength. We usually need to grind a bit on the inside of the band also, to keep it nice and smooth.

Fitting the band

It's best to have help for this process. One person should operate the torch (Figure 11.33) and another stand by with welding gloves on, ready to fit the band as soon as it's heated. Use an oxy-acetylene torch (a normal propane torch is not hot enough) to heat the band evenly by moving the torch head around the band. If you stop or go too slow it's easy to overheat or even burn through the band, so keep moving! After a few seconds the stainless band will just start turning slightly yellowish in color, at which time the other person should grab the band and quickly set it over the magnet rotor. It should be a very close fit—and then as the band cools it will shrink and fit very tightly around the rotor. Be careful as it cools!

Figure 11.33—Heating the band with an oxy-acetylene torch to expand it for fitting around the magnet rotor.

If the weld is weak or the band was cut too short, it is possible the band could break which could be dangerous, especially if somebody's face was too close. Safety goggles are recommended for this process.

Figure 11.34—Magnet rotor with the band shrunk in place. Note that this rotor uses larger, round magnets. We discuss this design modification in Chapter 12, *Stator*. Banding can be used no matter what shape your magnets are, though.

After the band has shrunk to fit, turn the rotor over and if necessary tap the band down on a flat surface so that it's flush with the front of the magnet rotor. Take a 6 inch diameter disk island, wax the sides and stick it to the center of the magnet rotor with caulk to create a "dam" in preparation for casting resin around the magnets. Put the magnet rotors on a level surface and pour the resin into the rotor up to the tops of the magnets and flush with the stainless steel band. If you accidentally slop resin on the tops of the magnets, remove it with a rag before it hardens.

Which method to choose?

As mentioned earlier, this method of making magnet rotors does leave the surface of the magnets exposed. It is possible that over time, (especially in a humid or saltwater climate) that they could corrode even though the nickel plate on the magnets should prevent that. If you are concerned about corrosion, we suggest a high quality epoxy paint over the inside of the magnet rotor and over the tops of the magnets.

Both the usual way to make magnet rotors and our new method here work fine, and both have their pros and cons. There are certainly other ways to fasten the magnets to the rotors as well--we have seen people acquire magnets with little holes in them and screw them down, or pin them down as well and avoid casting resin altogether. So long as the magnets are well fastened to the disks for the long term and some measure is taken to assure the magnets can never fly out, it should work fine. In building your own wind turbine you'll make the best choice based on your intuition and your resources.

Figure 11.35—The back magnet rotor in place, after banding and casting with the new method shown in this section.

12. Stator

The stator is a casting that contains and pro-tects all nine coils of wire across which the magnets spin. It must be a solid casting or else magnetic forces would make the coils (and the individual wires inside) move around, fatigue, and break. It's called a "stator" because it is the stationary part of the alternator—it doesn't spin. This is a three-phase alternator. Each phase will consist of three coils in series, with a "star" connection (Chapter 4, *Electricity From a spinning shaft*) in the center that's common to all three phases.

Figure 12.1—Completed stator.

You'll build and wire this together in the stator mold before casting. The lines you drew on the bottom of the stator mold when you built it will assure that the coils are the correct size and are placed correctly. A completed stator is shown at right in Figure 12.1.

Stator materials list

- 6 pounds magnet wire, double insulated, 200° C rated
- 15 inch diameter disc of fiberglass cloth or mat, quantity 2
- Thin viscosity cyanoacrylate glue (Superglue®) and accelerator
- 1/2 gallon resin
- Hardener catalyst for resin
- Disposable plastic mixing cups, 1 quart
- Plastic stirring rod
- Rosin-core solder (electrical type)
- Electrical tape
- Duct tape
- Heat shrink tubing
- Mold release compound (automotive or wood wax)
- 1/2 inch diameter bolt or dowel, 8 inches long

- Brass 1/4 – 20 tpi screws, 1-1/4 inches long, quantity 3
- Brass or copper 1/4 inch washers, quantity 12
- Brass 1/4 – 20 tpi nuts, quantity 6

Winding the coils

The magnet wire size to use for your coils depends on the voltage of your power system. Any design variations you choose to implement will affect your choices, too—be sure to read the *A more powerful alternator?* sidebar in Chapter 12 (*Stator*) before purchasing wire. Roughly speaking, every time you go up three AWG gauge sizes the wire has half the cross-sectional area. The voltage of the machine is directly related to the number of turns in the coils. If you double the number of turns in each coil, then the voltage is doubled. But no matter what the voltage of the machine, the coil size and total weight of copper need to remain about the same so they will fit into the space allotted. 12 volt machines require very thick wire—so thick that it's difficult to wind. To ease this process when winding 12 volt stators, you can use two spools of wire that's half the cross-sectional area and handle the two strands as if they were one while winding. In Figure 12.2, "volts" refers to the system voltage of your off-grid power system, "# AWG" is the wire gauge and "# turns" is how many loops you'll wind into each coil. If you followed the chapter on building your coil winder carefully, the coils should fit nicely in their allotted spaces in the stator mold. The whole stator will require about 6 pounds of wire.

Volts	#AWG	# Turns
12	2 strands #14	36
24	1 strand #14	70
48	1 strand #17	140

Figure 12.2—Wire sizes and number of turns for different system voltages.

The wire sizes and numbers of turns in Figure 12.2 are appropriate only if the stator is used with the magnet rotors and 10-foot blade set that we specify in this book! Small changes won't make much difference, but if you use different magnets, spacings or blade sizes the turbine won't perform correctly. Refer to our *Scaling it up and down* chapter for more information before you start tinkering around with changing the stator design. If in the future you upgrade to a higher voltage power system, you'll simply have to wind and cast a new wind turbine stator to replace your old one. It's a relatively simple procedure to swap out stators, and makes for an easy and inexpensive upgrade. If you will be experimenting with building the stator for a special application like a grid-tied system or an MPPT controller, be sure to study the sidebar on the next page, *Winding for grid tie or MPPT?*

Winding for grid tie or MPPT?

Grid-tied systems may require the stator to be wound for different voltages than the typical ones listed here. We are located 12 miles off grid, and so have never tried grid-tying any of our turbines—though other experimenters have grid-tied this design. And although MPPT wind turbine controllers are not yet on the market (see Chapter 4, *Electricity from a spinning shaft*) as of this writing, they may become popular in the future and will also require special considerations in winding the stator. Here are some brief points to keep in mind if you are planning a grid-tied or MPPT application:

• If the grid-tie system incorporates a battery bank (called an "islanding" system), no changes are needed. Wind the stator to match the battery bank voltage.

• If no battery bank is used (a "non-islanding" system) or if an MPPT wind controller will be used, consult the inverter or controller manufacturer to find out what voltage range the device wants to "see" from the wind turbine at the input terminals. The homebrew turbine's output will be very similar in form to other small commercial three-phase turbines that are similar in size. But you may have to increase the stator's designed voltage to match whatever direct grid inverter you have. And, the inverter itself may have to be custom programmed to match your turbine's power output curve.

• Doubling the number of turns doubles the voltage at any rpm, and using wire that's three sizes smaller lets you fit those extra turns in the same size coil. The coils must still fit in the same space in the stator, so you'll be using more turns of thinner wire. See Chapter 20, *Scaling it up and down*, for more information.

Clamp your new coil winding tool to a clear workbench about three to four feet from your bench vise. Scrounge a 1/2 inch diameter bolt or dowel 8 inches or more long from your scrap pile, and clamp it vertically in the vise. Place the magnet wire spool over the bolt. Clamp the coil winder to the workbench with a C clamp. The setup is shown in Figure 12.3.

Prepare your equipment and place it near the coil winder—you won't have any free hands once you start. You'll need a large pair of side cutters to cut the wire, and some electrical tape to bind the newly-wound coils together. If your coil winder has a wing nut holding it together, you can use your fingers to take it apart. Otherwise you'll need a 1/2 inch wrench. When you bolt the front onto the coil winder, finger-tight is fine. But after all the wire is wound into a coil the nut will be tight, so if you don't have a wing nut you'll need the wrench to get it apart.

Figure 12.3—Coil winder and wire spool, ready for action.

Figure 12.4—Bend in the wire.

Figure 12.5—Securing the wire in the coil winder slot and around the nut.

Figure 12.6—George demonstrating proper coil winding technique.

About 10 inches from the end of the wire, bend it sharply 90 degrees with the needlenose pliers as shown in Figure 12.4 at left. Drop the wire in the slot of the coil winder and bend the end around the nut like in Figure 12.5, below left. This will keep the wire from slipping when you wind the coil. Hold the wire tightly in one hand, keeping tension on it, while turning the crank with your other hand. Be careful to keep constant tension on the wire, and try to turn the coil winder at a constant speed. We notice that lots of folks tend to turn it faster on the down stroke and slower on the up stroke, and this often results in a lopsided coil (with one side of the coil wider than the other). It's important to keep constant speed and consistent tension on the wire while winding. Try to wind neatly, but don't obsess over perfection. We've seen some folks take over an hour to wind a single coil, trying to pile the wire in perfectly. This usually results in lopsided, sloppy coils. Relax. It should take no more than a minute or two to wind a coil—maybe five or ten if you are overly obsessive!

In Figure 12.6 at left, George is demonstrating proper coil-winding technique. Note that he is not stressing out on trying to keep every winding perfect, but rather keeping up a steady, rapid rhythm that places each winding in perfect alignment. After each coil, George enjoys a cool, relaxing beverage to calm his nerves for the next coil.

Once you have the correct number of turns in the coil, pull the wire out of the slot and twist the two ends together (a half twist, just enough to hold them together). Don't twist more than you need, because later you'll untwist this and it's not good to bend up the wire more than necessary—it will break after too much fatigue. Grab the wire between the spool and the coil with one hand, and clip it off so that there's about 10 inches of wire coming out of the coil. Now both leads coming from the coil should be

Figure 12.7—Freshly-wound coil, still on the winder pins.

about 10 inches long. Take the loose end that's coming from the spool of wire, put it on the workbench and set something heavy on it (the side cutters are handy, since they're probably still in your hand) so that the rest of the wire on the spool doesn't unravel spon-

Figure 12.8—A completed, taped-up coil.

taneously. Take off the end of the coil winder. The coil will come with it. The coil should pretty much fall off the end if you just turn it over. Do it carefully so the coil doesn't fall apart, Figure 12.7 above left. We refer to the longest two sides of the coil as the "legs." Tape the legs of the coils with a couple wraps of electrical tape to hold things together, Figure 12.8 above right.

One coil is finished, with eight more to go!

Check each coil after you tape it, each should fit in the stator mold between the radial marks, as shown in Figure 12.9 at right. It's OK if they're a tiny bit smaller. If you made a 48 volt stator with #17 wire, they may be slightly undersized—that's not a problem. Remember that the stator mold has nine radial lines that tell you the maximum width of the coil, and it has two circles (8 inch and 12 inch diameter) that show you the path of the moving magnets. When checking the size of the coil, the hole in the coil should center over the 8 inch and 12 inch circles. In that position the coil will fit in between two of the radial lines. If the first coil fits well, then wind eight more just like it! In Figure 12.10 all nine finished coils are in the mold, and you can see how they are supposed to fit. Again: It's OK if they come out slightly smaller than those pictured, but any

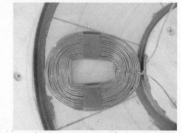

Figure 12.9—A finished coil placed in the stator mold to check its size.

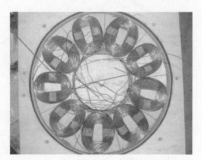

Figure 12.10—All nine coils placed in the stator mold.

larger is a problem—if they are too big, it means the coils were wound too loosely and you'll have to re-wind them.

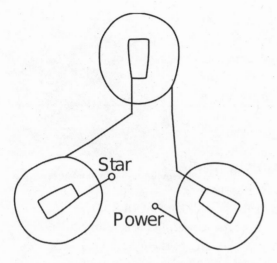

Figure 12.11—Wiring diagram for one phase. Each of the three phases is wired the same way.

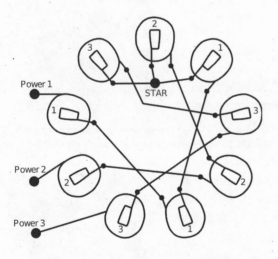

Figure 12.12—Wiring diagram, all three phases.

Wiring the Stator

Figure 12.11 at left shows how the first three coils are connected together into one phase. Each phase is numbered and consists of three coils in series. We define each coil (and each phase) as having a "start" and an "end." The start is the lead that comes from the inside of a coil, and the end is the lead from the outside of a coil. In the final product, the common "star" connection is made near the inner diameter of the circle—the starts from each phase are connected together, and this connection is permanently encapsulated in the resin. The series connections for each phase are also embedded in the resin, on the inside of the ring. Figure 12.12, below left, shows all three phases and how they are wired together.

Take three coils and put them in the mold in their proper position. Pick three spaces which are 120 degrees apart. If you could superimpose an image of the magnet rotors over these three coils you'd notice that at any given time, each coil is seeing an identical "magnetic situation," and therefore they are "in phase" with one another—when one of them is at maximum voltage, the other two will be also. Be sure all three coils are placed same side up, so that the start of the coil is the wire that crosses over the inside, facing up. You need to be sure that no coils are upside down.

Take the "end" of one coil and wrap it around so that it points towards the outside of the mold. Tape it to the leg of the coil. Basically you are adding half of a turn to the coil when you do this. It used to point towards the inside of the mold, now it should point to the outside. Then take the "start" (the wire coming from the inside of the coil) of the same coil and bend it around the island in the middle of the mold to the next radial line in the mold, and cut it off about 1/2 inch past that line. This wire will connect to the "end" wire of the next coil, and the "start" of that coil will need to connect to the "end" of the last coil in that phase. What you are doing here is making sure you have just the right amount of wire to connect the coils to each other, and cutting off the excess. It's better to be a bit on the long side than a bit on the short side, so give yourself a little extra length to be sure—but not much because there's not much room for extra wire on the inside of the mold. Once this is done to three coils, you have the first phase almost ready to solder together, as shown in Figure 12.13. You'll be repeating this for the next two phases.

Figure 12.13—One phase, almost ready to solder.

The enamel insulation on good quality magnet wire is fairly thick and very hard to scrape off. It's usually double insulated and the inside layer of insulation is almost invisible, so even though you might think you've done a good job of stripping the wire, it could still be on there. It can be deceiving! The easiest way to assure complete stripping is to burn the insulation with a propane torch, about one inch back from the end of the wire. We usually heat it enough so that the wire itself becomes red hot. This also anneals the copper and makes it easier to twist together. Do this to all the leads that you've cut. You've not cut the "end" of the first coil yet (it sticks out towards the outside of the mold) and you've not cut the inside (the "start") of the last coil yet, so leave those be for now—you'll deal with them later. Let the wires cool down, and then clean the burned insulation off carefully with sandpaper.

Put the three coils back in the mold and twist the stripped wire ends together tightly with needlenose pliers. If you measured right when clipping them off, the connections should be pretty much centered in between the coils. Now solder the connections, bend them over with pliers, and insulate with electrical tape, keeping everything as thin as possible. You can also use heat shrink tubing here for neater appearance—if you do, be sure to put that on before you twist the leads together! If you're building a 12 volt machine with multiple strands of wire in each coil, it can be tricky to twist things together tightly. You can use short pieces of small copper or brass tubing as "butt splices," then crimp them tightly and solder as shown in Figure 12.14. Once

Figure 12.14—A butt splice connection.

you have one phase soldered together, carefully remove it from the mold and do the same thing to the other two phases.

After all the phases are finished, put all three back into the mold. Do it such that your three "ends" (those wires that are pointing towards the outside of the mold) are beside each other. Those three "ends" will be the power leads out of the stator (the output from the wind turbine). Caution is needed at this point! Be sure these connections lay in the mold neatly and are not jammed up on top of other wires—when the mold lid goes on, it will be clamped down tight and could compress a jumble of wires and butt splices enough to damage the insulation or electrical tape, causing a short circuit.

Tips for solid soldering

A bad solder connection is worse than no solder at all! And thick wires are more difficult to solder than thin ones. The stator wires will be carrying lots of current, so a good connection is critical. Here are some tips for proper soldering technique:

• Make sure all wires are completely stripped of insulation, and are clean and shiny. Remove any oxidation (dull color) with sandpaper.

• Use a powerful soldering iron of at least 100 watts, or a soldering iron tip on a propane torch. Don't use just a torch flame, it oxidizes the outer layer of copper as you heat it.

• Use rosin-core electronics solder, not plumbing solder.

• Heat the wires, not the solder. Get them to proper temperature before applying solder.

• Continue heating the wires as the solder flows in, and make sure everything is wetted with it.

• Don't jostle the joint as you remove the soldering iron; let the solder cool for a bit before moving anything.

Next you need to make the star connection between the phases. The three inside leads should be brought together on the inner diameter of the stator so that you can connect them together. Leave enough slack in the wires so that you can make this connection, and then push it down between the coils and the island in the mold. So, once again—figure the length, cut the wires off, burn the insulation with the torch, sand it, twist them together and solder them. Then insulate the connection with electrical tape or heat shrink. Figure 12.15 below shows the finished star connection. All that needs to be done now is to poke it down so it doesn't stick up above the coils. Again, make sure it sits in there neatly so nothing will be damaged when the lid mashes everything down tight.

Figure 12.15—The star connection.

Even if the coils fit well and are of the right size, once you make all these solder connections things are kind of springy and the coils will not stay in their proper positions. The connections you made and all that wire around the inside is sure to push some of the coils out further than they ought to be. Use duct tape and go around the stator one coil at a time, taping each coil down exactly where it belongs between the marks on the stator mold. Keep the tape off the "legs" of the coils as shown in Figure 12.16. Then cut nine little rectangles from fiberglass cloth, about 1-1/2 inches wide and 2 inches long. Before you cast the stator you'll need to remove the coils from the mold again, and you will use these rectangles of cloth and cyanoacrylate glue to hold everything together so that you can move all the coils together as one piece.

Figure 12.16—Taping the coils into position.

Glue the fabric rectangles to the legs of the coils with cyanoacrylate, as shown in Figure 12.17 at right. Put plenty in—you'll know it's a good glue joint when the cloth becomes transparent. This will also serve to pot the legs of the coils and prevent individual wires from vibrating against one another inside. It might not actually be a big issue, but we like to have lots of glue inside the coils. Also put glue on the fabric between the coils so that the fabric becomes hard and less flexible. This will make the stator wiring fairly rigid and easy to handle before you cast it. Try hard not to glue the coils to the mold! Remove the duct tape, it should pull off easily. Now you can carefully pick up the stator coils as one unit, and remove it from the mold. It should look like Figure 12.18, below left. Put it somewhere safe until you're ready to cast the stator in resin.

Figure 12.17—Fiberglass squares glued to the coils to hold them in place.

Figure 12.18—Coils glued together and removed from the mold.

Casting the stator

The stator will be cast with the same resin you decided upon and used for the magnet rotors, and may or may not include fillers. Review the *Magnet rotors* chapter and *Resins* sidebar before doing this, as the procedure is very similar and the safety precautions are the same. You'll need a ring of fiberglass cloth on both sides of the coils. If you fold the fabric correctly (Figure 12.19), you can lay out just one quarter of the ring and cut out two rings in one shot, Figure 12.20 at left. The rings are 15 inches in outer diameter and 6 inches in inner diameter. Set them and the coils aside in a safe place for now, while you prepare for casting.

Figure 12.19—Fold the fiberglass cloth into quarters for easy layout.

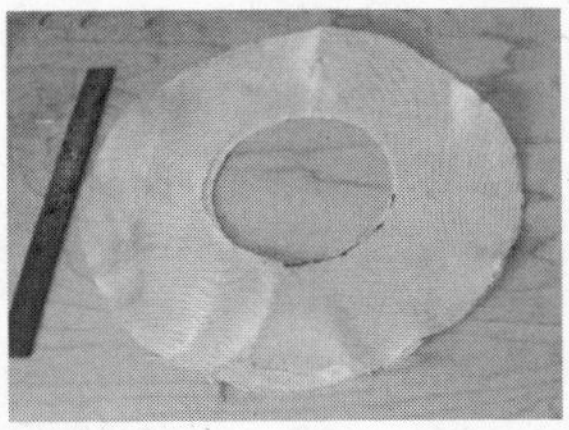

Figure 12.20—Fiberglass ring ready for casting. You'll need two of them.

Grease the stator mold inside, on top, and around the inner edge. Same with the lid—everything that will touch the resin should be thoroughly covered with grease or wax as a mold release compound. Automotive or wood wax works quite well. In a pinch (it's a long drive to the hardware store from here) we've used axle grease, crayons or vegetable shortening! None of these work as well as wood or car wax, but they do work.

It takes just about 1/2 gallon of resin and filler to cast the stator. Find a level place to put the mold, with a surface that won't be damaged by spillage. It's important to either keep the edges of the mold off the workbench, or find a work bench that allows access for C clamps around the edges of the mold. After the resin is poured you'll need to clamp the lid down with C clamps, so think ahead!

Get a one gallon plastic mixing cup (or a milk jug with the top cut off) and a sturdy stick of plastic to stir with. The plastic dividers that your magnets were packed with work perfectly for this. Put on your protective goggles and gloves (and respirator if you are working inside). Pour about a quart of resin into a mixing cup, add the filler (if any), and mix thoroughly. Add more resin and filler to make 1/2 gallon. Add the hardener to the resin, and again mix thoroughly. Too little hardener and the casting could take days to set, if ever. Too much and it will set before you get the coils in there. *Way* too much and you'll have a time bomb on your hands that could get so hot it melts the plastic cup or catches on fire. We've been there and done that! Mix the resin and the hardener thoroughly with the stir stick.

Pour about a pint of resin into the mold and roll the mold around so that the whole bottom and the sides become covered with resin. Put one of the fiberglass rings into the resin (Figure 12.21 at right) and work it with your stir stick until it becomes saturated. When saturated the fabric will become almost invisible and you won't see any white. Then pour more resin in the mold, and work it into the fabric more—try to work out any air bubbles.

Figure 12.21—Working the resin into the fiberglass.

Put the entire coil assembly into the mold carefully (Figure 12.22 below left) and poke gently at everything with the stir stick so resin runs all around the coils and any air bubbles rise to the surface. The ends of the wires must protrude out of the mold. Then fill the mold the rest of the way with resin and be sure all surfaces of the coils have resin on them. Put the remaining fiberglass ring over the top of the coils and work resin into it—again, it should almost disappear. Pour the remaining resin over the top of the fabric. Work it in and again try to work out air bubbles. It might not hurt at this time to beat on the mold a bit or vibrate it with a sander (or something) for a couple minutes to help air bubbles rise to the top.

Figure 12.22—Putting the coils into the goo and distributing the resin.

Carefully put the lid of the mold down over the casting. Put a 1/2 inch washer over the threaded rod, and run the 1/2 inch nut down over it. Tighten the nut—this does a good job of clamping the lid tightly on the mold and ensuring that the finished casting will be exactly 1/2 inch thick. Put a C clamp on each side of the mold (use four C clamps) and tighten them evenly, as shown in Figure 12.23 at right.

You'll have some idea when the resin is hard inside from observing all the goo that spilled out the side! Let the stator sit in the mold until the resin is hard. A chisel works well to scrape and peel the resin off from

Figure 12.23—Tightening the mold lid and clamping the edges.

around the outside of the mold. A good time to do this is while it's still a bit flexible (before it gets really hard). Once the resin has hardened completely, remove the C clamps and the nut in the center. Use a chisel or a screwdriver to gently pry around the lid until it breaks loose. Once the lid comes off you can usually turn the mold upside down and the stator will just fall out. If not, then turn the mold upside down and tap on it with a hammer or pry gently at the edge of the stator. It should come out easily if you greased the mold sufficiently.

Figure 12.24—George cleaning up the edges.

Use a file or a sander (or both) to clean up the edges of the stator, as demonstrated by George in Figure 12.24 at left. The inside diameter must be pretty clean because there's not a lot of extra room between this hole and the wheel hub which will rotate inside it, but be careful not to hit copper while cleaning it up—the wires are very close to the edge of the casting. Next, you'll need to drill the three holes for the studs that hold the stator to the stator bracket and wind turbine frame. Clamp the stator bracket to the stator on center as shown in Figure 12.25 at right and drill 1/2 inch holes

Figure 12.25—Marking the drilling pattern for stator bracket holes.

right through it. You don't want to hit any copper inside, that would ruin the stator! It's also very important that the center hole of the stator be perfectly centered with the stator bracket.

Next, drill the three holes for the power connection terminals. Be sure to have the holes come out between coils—preferably with the three wires coming out between the same two holes, as shown from both front (Figure 12.26, at left) and back (Figure 12.27, facing page) in the photos. Again, you don't want to hit

Figure 12.26—Completed stator, front view.

copper with the drill bit! If you do, you'll have to throw the stator away and rebuild and re-cast the entire thing. Drill the three holes 1/4 inch in diameter about 3/4 inch away from the edge of the stator—one near each of the leads that's coming out. Insert a brass 1/4 inch – 20 tpi screw through each one, with a washer on each side and a nut on the back. It's essential to use brass hardware here, as ferrous terminal bolts will cause magnetic drag in the alternator. Cut the wire ends just

Figure 12.27—Completed stator, back side view.

long enough so you can clamp each one between the bolt head and the washer. You can put a couple more washers and one nut on each screw now. These will serve as the power lugs to which you will eventually connect the power output lines to the alternator.

The stator is now finished! You can put it aside until you are ready to assemble the alternator.

A more powerful alternator?

The instructions in this book lay out how to build the magnet rotors with 1 inch x 2 inch x 1/2 inch Nd-FeB magnets, and explain exactly what gauge of wire to use and how many turns are required per coil in the stator. However, we also often build the same 10-foot machine with larger magnets, 2 inch diameter x 1/2 inch thick discs, and a stator that is wound with much lower resistance. The drawback of doing this is a higher cost in magnets, and the benefit is the lower resistance. You might consider this alternative if:

• You live on a very windy site with high sustained winds

• Your tower is very distant from the batteries

• You have a 12 volt power system

Let's explain those situations a bit more.

First, at very windy sites, we have on occasion seen stators overheat and burn out when we used the magnets and the stator as described in these plans. In most cases, though, the burnouts could have been prevented by making the machine furl earlier.

Next, if you are dealing with a turbine site that's more than 200 feet away from your battery bank, then odds are you'll either spend a ton of money on copper wire to transfer power from the turbine to the batteries, or you'll have lots of resistance in the line (which means lots of power lost as heat in the line). To the wind turbine blades, it doesn't matter if there is resistance in the line or resistance in the stator—they "see" the same thing. Resistance anywhere in the circuit will result in power loss and will usually cause the blades to run faster. If the blades run too fast they may get noisy, and the wind turbine won't start to furl until the winds are higher. Both of these problems are hard on the machine.

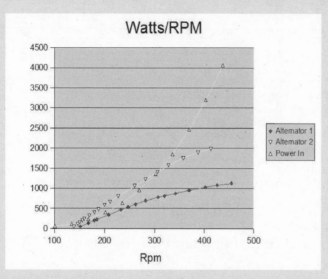

Figure 12.28—The power output curves of the alternator design in this book (alternator 1), and the more powerful alternator from this experiment (alternator 2), compared to the power output and rpm that could theoretically come in from the wind.

Lastly, if your system runs on 12 volts, you have about the same problem as a distant tower. Due to the low voltage, a 12 volt system requires very heavy cable to handle the high current. Generally losses in all parts of 12 volt systems are high. To make up for it, a stator with much lower resistance will help increase power gained from the turbine.

The chart at left (Figure 12.28) shows a few things: Power coming in (this is the power available at the shaft from the blades) is the steep line that goes to the top of the graph. It represents the theoretical power you could get from a "perfect" 10-foot rotor running at a tip speed ratio of 6. The actual power input will be less, and will depend on the quality of the blade set you carved.

The power curves for alternators 1 and 2 were measured using a tractor PTO to power them, a tachometer, and a multimeter (Figure 12.29 at left). The alternators were charging a battery bank that was held at a constant voltage. We sometimes even run the alternators so hard that they burst into flames, just so we can find their upper limits! It doesn't help that with tractor testing, there's no wind blowing across the stator to cool it.

Figure 12.29—Destructive alternator testing using a tractor PTO.

Alternator 1 is the lowest line on the chart and shows wattage versus rpm from the alternator built with 1 inch x 2 inch x 1/2 inch block magnets as described in the plans. Alternator 2 in the chart is built with the larger 2 inch diameter x 1/2 inch thick disk magnets shown here in this sidebar, and the stator is wound with thicker wire and fewer turns per coil. It has much lower resistance (about half) than alternator 1, and therefore at any power output level the stator needs to dissipate about half the heat! If you look at the chart at any given rpm, the difference between power in, and power out could be considered heat in the stator. We find that in most conditions (it varies with weather and wind speed) this stator as described cannot dissipate more than 700 watts continuously.

If you decide to go with this slightly different design with the larger magnets, you'll have to make changes in the stator too. However, the only modifications that need to be made are in the wire size and number of turns. We talk about this a bit more in the "*Scaling it up and down*" chapter, but you can see how small the differences are. That's one reason that designing an alternator to match a certain set of blades is a bit of a "black art."

Stator modifications needed for this More Powerful Alternator:

• 12 volt system—29 turns per coil using two strands of #13 AWG wire in hand.

• 24 volt system—55 turns per coil using two strands of #15 AWG wire in hand.

• 48 volt system—105 turns per coil using a single strand of #15 AWG wire in hand.

Adjusting the resistance

One advantage of this more powerful alternator design is that it lets you change the line resistance to "ease into" the proper performance level for your system. If it turns out that your situation is not so extreme as you thought with high winds, long wire runs, or low system voltage, you'll find that the larger magnets in this More Powerful Alternator can overpower the blades and stall them if the resistance in the rest of the system is too low.

In some cases (depending on battery voltage and resistance in the line), you may need to add resistance to the system to keep the blades running at a faster, more appropriate speed. You'll get an idea if this is necessary by watching both the turbine and your meters while the machine is making power. If the turbine cuts in nicely in low winds, but fails to speed up and generate good power in higher winds, then the blades are stalling—in that case, adding some resistance to the system would be appropriate. We usually add resistance between the rectifier and the batteries.

It's best to start out small (a little bit of resistance) and then add more as needed. This is all very easy, since you can do it from the ground with your wiring and watch the results high up on the tower top. A good way to experiment is to start by adding a fairly thin (#12 or #14 AWG) section of wire to the line. Use a wire gauge chart that shows resistance per given length, and work out exactly how many ohms you are adding.

It can take a bit of experimentation to get optimal performance from the wind turbine. We can't tell

Figure 12.30—A banded magnet rotor from this more powerful machine, using the larger round magnets described here and in Chapter 11, *Magnet rotors.*

you exactly how much resistance to add, but we can give some suggestions about the *most* you would want to add:

• 48V: Not more than 4 ohms

• 24V: Not more than 1 ohm

• 12V: Not more than 1/4 ohm

Adjusting the air gap

Another adjustment that can be made to the alternator is the air gap, which is defined as the distance between the magnet rotors—it is the space in which the stator is mounted. By increasing the air gap the magnet flux through the coils is reduced. Since voltage is directly related to magnetic flux, doing this has the effect of increasing the cut-in speed of the wind turbine, and flattening the power curve. Adding resistance (as discussed earlier) does not affect cut-in speed, but it does flatten the power curve. By flattening the power curve we mean that power out will be reduced over the full operating range of the alternator.

When we use larger magnets and coils as described earlier, the cut-in speed is slightly too low—so we usually add between 0.060 inch and 0.125 inch more to the air gap to bring the cut-in speed up to the desired 140 rpm. This is accomplished by adding one or two 1/2 inch AN washers on each stud, between the hub and the front magnet rotor. AN washers are thinner, and have smaller outer diameter than "standard" washers. If this seems confusing now, read ahead to Chapter 15, *Alternator assembly*, and it should become clear.

The other benefit of the wider air gap is additional mechanical clearance between the magnet rotors and the stator, which reduces the risk of them rubbing together. The alternator built with the larger magnets is basically a bit too powerful for 10-foot diameter blades. By adjusting the air gap and adding appropriate resistance to the line, you can "tune" the alternator to be a good match—and in the end you wind up with a cooler running, more reliable machine. The drawbacks of going this route are higher magnet cost, and some time spent fiddling around matching the alternator to the blades. If you use the larger magnets and wind the coils as suggested in this section, then a wider air gap is appropriate.

Good luck, and have fun experimenting with "tuning" your new wind turbine to perfection! Also take a look at Chapter 20, *Scaling it up and down*, for more alternator adjustment information.

DOG HAIKU

Engineers design
But rarely build a turbine
Do they have no thumbs?

13. Rectifiers

The alternator you're building produces three-phase alternating current. In order to charge batteries you need direct current. You can easily accomplish this conversion by running the alternator's output through an array of rectifiers. The terminology here is a bit confusing—rectifiers are also called diodes, but what we need is an array of diodes. Two diodes in a circuit can be wired into what's called a "half-wave bridge" rectifier, and four diodes make a "full-wave bridge" rectifier. Full-wave bridges are a very common and inexpensive electronics component, so it's much simpler, cheaper and neater to simply use three or six of these than to make your own bridges out of individual diodes!

Figure 13.1—A typical 35 amp, full-wave bridge rectifier.

You'll be using one full-wave bridge rectifier on each phase of the alternator output (Figure 13.1). For this project you'll be doubling up both sides of three single phase full wave bridges, making a three-phase bridge. The wiring for this is shown in Figure 13.2. The standard full-wave bridge that's commonly available at electronics stores is rated at 35 amps. That's ample current-carrying capacity for a 48 volt machine, but too much for 12 volt or 24 volt alternators—in

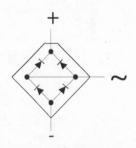

Figure 13.2—The diodes inside a typical 35 amp, full-wave bridge rectifier.

these cases you'll need six rectifiers and will wire them in parallel pairs for more capacity. It's possible to find larger ones, but prices go up exponentially above a 35 amp rating, so it's cheaper just to use multiple 35 amp models. If you can find a pre-made, three-phase rectifier at a reasonable cost, go with it! These are often available on Ebay or at surplus electronics outlets at half price or less. It will eliminate many of the wiring steps described below. No matter which type you choose, you'll need a total of 40 amps capacity for a 48 volt system, 80 amps at 24 volts, or 120 amps for a 12 volt system.

It's handy to have the wind turbine shutdown switch indoors and near the rest of your power system, so we build it right into the rectifier assembly. It's also nice to have an ammeter to show your turbine's output at any given time, and the rectifier assembly is again a convenient place to mount this. Both the switch and the meter could be mounted elsewhere in the system if you prefer.

Materials list

- Full-wave bridge rectifiers, minimum 35 amp rating, quantity 3 or 6

- Mounting screws for each rectifier, usually #10 – 24 tpi.

- Aluminum heat sink

- Terminal blocks, quantity 3

- Silicone heat sink compound

- Spade connectors for 10 gauge wire, quantity 12

- 10 gauge insulated stranded wire, 6 feet

- 3PST (3 pole single throw) switch, minimum 35 amp rating

- Ammeter, 40, 80 or 160 amp (see text, optional)

The parts

The heat sink is simply a large piece of finned aluminum. This can be fairly expensive to buy (see our *Sources* chapter), but it can also easily be salvaged from old electronic equipment. You'll find nice big ones on old car ste-

Figure 13.3—Heat sink, top view.

reo power amplifiers, power inverters and lots of other places. We've used finned aluminum cylinder heads from small engines as well. The heat sink draws heat off the rectifiers and dissipates it to the air. There is approximately a 1.4 volt drop across the bridge rectifiers. You can multiply that number by the current flowing through them to figure out how much heat will be generated in the rectifiers. For example: if the wind turbine is producing 10 amps, then 14 watts will be wasted as heat at the rectifiers.

Figure 13.3—Heat sink, side view.

Without a heat sink, heat will build up quickly in the rectifiers until they overheat and fail. The lower the system voltage, the higher the current through the rectifiers will be so a larger heat sink is required. In 12 volt systems about 10 percent of the energy produced by the wind turbine is wasted as heat, and a large heat sink is required (consider that if the wind turbine is producing 100 amps then your rectifiers are a 140 watt heater). This is one of many reasons to avoid lower voltage systems. It's a good idea when you finish your homebrew wind turbine system to watch the rectifiers closely to see how hot they get in high winds. If they seem to be getting too hot to touch, rebuild the rectifier assembly on a larger heat sink.

Each full-wave bridge rectifier has four leads (Figure 13.1). Two leads (on opposite corners) accept incoming alternating current, and the other two put direct current out. Sometimes all four leads are marked, but in most cases only the positive (+) DC lead is marked. Usually the positive spade terminal will have a flat corner near it, and the spade is usually at a 90 degree angle to all the other terminals. Negative is on the opposite corner from positive.

The schematic diagram of what you're building is shown at right in Figure 13.4. You need three rectifiers and they should be rated for the full current you expect from the wind turbine. In reality each rectifier must actually handle about two thirds of the total current, but if you rate them for at least the full current (and preferably a bit more) that your wind turbine will ever produce then you have a bit of a safety factor. For this 10-foot diameter machine, figure maximum output to be around 1,200 watts. At 48 volts that would be about 30 amps so you can use three 35 amp bridge rectifiers safely. At lower voltages you'll need heavier rectifiers—or, you can wire up multiple rectifiers in parallel.

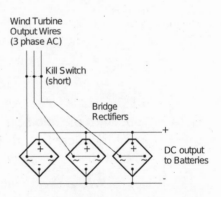

Figure 13.4—Schematic of the rectifier array you are building.

Each output wire from the wind turbine will connect to both AC leads of its own bridge rectifier. The DC leads of the three rectifiers are in parallel and will be wired directly to the battery later. The schematic also shows the turbine shutdown switch if you choose to mount it on the rectifier assembly (Figure 13.4). All it does is short all three of the AC leads together when the switch is closed, and leaves them all as-is when the switch is open. This will stop the turbine from spinning in just a few seconds.

We use terminal blocks salvaged from old electrical equipment, though you can buy them at an electrical supply shop for a hefty price. Be sure they are sized large enough so that one side will take the larger wire coming in from your wind turbine. The other side needs to take only the #10 AWG that will go to the rectifiers. #10 AWG is big enough for the output of a 10-foot turbine at

48 volts, and if you wire for 12 or 24 volts you'll be using multiple wires to multiple rectifiers—so, #10 is fine for these situations too. The terminal blocks are shown in Figure 13.5.

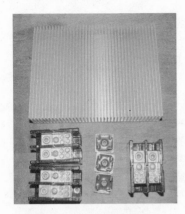

Figure 13.5—The heat sink, rectifiers (center) and terminal blocks.

Each rectifier will usually connect via what are called spade terminals (Figure 13.6)—these make wiring everything up very simple. You simply strip the insulation off the end of the wire, insert it into the spade terminal, and crimp it soundly with a crimping tool. Crimp connectors do not work with solid wire, only stranded! That's why we call for stranded wire, plus the fact that it's much easier to bend and work with.

Figure 13.6—A crimp-on spade terminal for stranded wire.

Your turbine shutdown switch should be a heavy-duty DPST (double pole single throw) model that's rated for at least 35 amps. If you shut down the turbine in high winds, the current through the switch could be a great deal more than that (especially in a 12 volt system) but it will only have to carry that for a second or two as the turbine slows down.

We are very fond of old analog meters, and that's what we always use with our wind turbines to show the output in amps. You'll need a meter that can show up to 40 amps if you are building a 48 volt turbine, 80 amps for a 24 volt turbine and 160 amps for a 12 volt turbine. If you can't find a meter that can handle enough current, it's not very difficult to modify a smaller one by adding an extra shunt, calculated with Ohm's law—this lets more current flow around the meter instead of through it.

Assembly instructions

Figure 13.7—Tapping the holes for mounting the rectifiers and terminal blocks to the heat sink.

The first step is to pick locations on your heat sink for the rectifiers and the terminal blocks. Drill the holes and tap them (Figure 13.7). Usually #10 – 24 tpi screws are appropriate for mounting the rectifiers. Put a bit of silicone heat sink compound on the bottom of each rectifier before you screw it down. It's a sort of grease that helps the heat sink draw heat off the rectifier more efficiently, and is essential.

Mount the rectifiers and the terminal blocks to the heat sink. The terminal blocks in Figure 13.8 are larger than necessary, but that sort of thing is typical when you're salvaging parts from other equipment. Better to have things oversized than undersized!

Figure 13.8—Terminal blocks and rectifiers mounted to the heat sink.

Figure 13.9—AC input wires connected from terminal blocks to rectifiers.

In Figure 13.9 at left, the AC leads of all three rectifiers are connected to the three terminal blocks which will connect to the wind turbine. Now you can simply connect all the positive DC (+) leads together and bring them to the DC terminal block (at the upper right side of the picture in Figure 13.9), and do the same with the negative DC (−) leads. One half of that terminal block will be the postive, the other half the negative.

The finished rectifier assembly is shown in Figure 13.10 at right. The one in the picture also has an ammeter in series with the negative DC line so that the output from the wind turbine can be monitored. The kill switch is wired into the AC end. It's nice to get all the components in one clean unit like this. Also, you should make some kind of bracket or box so that the rectifier assembly can be mounted to a wall. It should be mounted so that the heat sink is vertical, (not laying flat) and the fins should run vertically (*not* horizontally) so that air will flow up between them more efficiently—this improves their ability to transfer heat to the air.

Now that your rectifier system is complete, it's time to assemble the alternator and test it!

Figure 13.10—Completed rectifier assembly, with ammeter and shutdown switch installed.

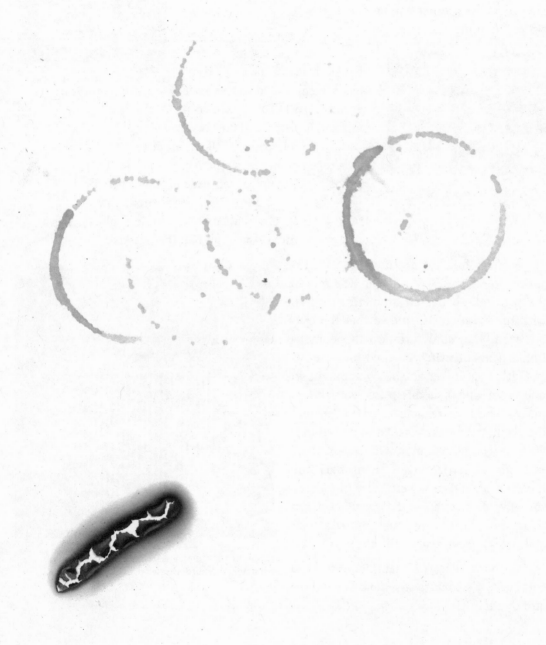

14. Alternator assembly

After you've built all your frame and alternator components, the next step is to clean and paint the metal ones. A good quality paint job can be important to protect the turbine from rust, though it is still optional—whether you paint or not depends on your climate, whether you have saltwater nearby and finally on your sense of aesthetics. You should paint all the metal parts separately, before you assemble the machine. This is also the time to finish up any grinding and deburring that remains. Clean the parts carefully before painting, and make sure any scale left on metal parts is removed by sanding, grinding or steel wool. The sidebar on the next page about painting will show you how we do it up here at our shop—we like to take the time to do an excellent paint job using high-quality, expensive automobile primer and enamel. It makes the machine look less like a homebrew garage experiment and more like a commercial wind turbine. We actually get many of our machines powder coated now, but that's not something you can do at home.

Figure 14.1—Completed frame on stand, ready for the alternator assembly.

After the paint has dried, it's time to assemble and test the alternator. Be sure that your shop is clean and ready—you'll need a clear work area free of ferrous objects that might snap to the magnet rotors, and all parts carefully checked for metal burrs or other imperfections that might hinder assembly. Sweep up all metal shavings from floors and workbenches, the magnet rotors will attract them too.

Please note that the finished magnet rotors can be extremely dangerous! They will ferociously attract any magnet or ferrous object from a long distance, and this could injure you or damage a magnet on the rotors. If the two rotors are allowed near each other, or are set on a steel workbench, your hand and the bones within could be crushed into a pulp, and it might require the fire department and extrication equipment to separate the rotors.

Figure 14.1 above shows the completed frame on its stand, and Figure 14.2 shows the assembled alternator after you've finished with this chapter.

Figure 14.2—Completed alternator.

Painting pointers

This section is by no means a complete guide to painting metal, nor are we anything close to painting experts! In fact, we get most of our turbines powder coated down in town these days. We simply want to give you some tips and cover some of the issues involved.

Safety—There's likely going to be sanding involved, so a dust mask should be worn. Solvents, paints and primers are almost all toxic and flammable. It's important to have good ventilation and a respirator. Every chemical you use will have instructions and safety information with it. Actually read that stuff and do as they say! If you wear glasses, cover them with protective goggles or you may never get the overspray off.

Preparation—Before painting it's important to prepare and clean all the parts. As they say, proper preparation prevents poor paint performance! Oftentimes the steel parts will have some scale and rust on them. A bit of rust is OK, but it's nice to remove as much as possible. Usually we'll go over the whole machine with 100 grit or finer sand paper. A vibrating or an orbital sander speeds this up quite a bit. Once we've sanded things to our satisfaction we clean all the surfaces that will be painted with a good solvent. Lacquer thinner works well for this—we just wipe all the surfaces with a rag soaked in solvent to be sure that all the grease, oil and fingerprints are gone. Paint doesn't stick to oil. After everything is cleaned up we then mask off any parts that we don't want painted. We normally mask off at least the spindle. If we choose to paint the wheel hub, we mask off the inside of it so we don't get paint on the bearing races.

Color choices—You, the builder of the machine, will probably believe it's a thing of beauty and you may want a paint job that really stands out. Neighbors may feel differently! Most commercially manufactured wind turbines are painted with light colors that blend in with the sky so that the machine creates as little visual impact as possible. White, light gray or light blue are common colors. Again though—if you're the only person who will see it, do as you see fit. Heat is one other consideration and it may be more or less important depending on your local climate. NdFeB magnets will start to lose their strength near 200 degrees F. This is a pretty low temperature and in some hot, sunny places you might have problems if you paint the magnet rotors dark or black. In high winds the stator can also get fairly hot, so again, a dark color may not be a good idea.

Paints and primers—We use a compressor, an HVLP (high volume low pressure) spray gun and automotive paint. A good "self-etching" primer designed for automotive painting is needed, and sticks pretty well even if there is a bit of rust left on the surface of the steel. After we prime the machine then we normally paint it with acrylic enamel. All of this stuff is available from automotive paint suppliers. If you go this route you'll need to read carefully about how the paints and primers should be mixed, what sorts of hardeners are required, the dry times, etc. It's important to follow the manufacturer's instructions carefully and make sure all the chemicals you use are compatible.

A simpler alternative is the "spray bomb"—canned spray paint. Lots of folks will take this route and it works fine. It won't hold up as long or be as glossy and smooth as a good acrylic enamel, but odds are nobody will notice once it's up on a tower! If you choose to use spray paint, first use the highest quality self-etching metal primer that you can find. Once the surface is primed and has dried for the correct amount of time, a good coat (or two or three) of high quality paint that's designed for use on metal should work fine. When using spray cans, we often choose "tractor and implement" paint, it's designed to hold up well outdoors on machinery.

A good paint job is sort of the final touch for your project. It may not serve much practical purpose except rust protection, but it can make the difference between a machine that looks "homebrewed" versus one that's been professionally manufactured.

Figure 14.3 at right shows the front magnet rotor. It has four extra holes for the jacking screws. Figure 14.4 below right shows the back magnet rotor. Note that when you built each rotor, you made an index mark both on the face and on the edge. These marks ensure that the magnetic polarity of the two rotors lines up correctly during final assembly. The jacking screws are used to gently lower the front magnet rotor into place. Without them, the ferocious attraction between the rotors would slam the front rotor into place, possibly causing damage to it or injury to you.

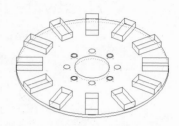

Figure 14.3—Front magnet rotor.

Figure 14.1, previous page, shows the assembled and painted frame all ready to take the alternator. It's very helpful to build a bench-top stand for assembly and testing, if you haven't built it already when you were welding the frame together. Ours is simply a 2 inch diameter pipe stub welded to a car wheel.

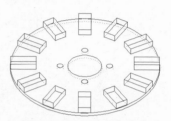

Figure 14.4—Back magnet rotor.

Pictured at right in Figure 14.5 are the wheel hub, bearings, washer, cotter pin, and nut. Typically when ordering the hub, you also buy the bearings. The washer, nut, and cotter pin come with the spindle. The hub is Dexter part number 81-9A and the spindle is designed for this hub, but they are always sold separately. Most trailer supply shops can provide both of these parts. We like the Dexter hub because it seems to be the most common one available, and it comes machined

Figure 14.5—Hub and bearings.

flat on both sides. If you use a different model of hub, the lengths of the all-thread studs will change, and you may have to use extra shims to set the air gap between the magnet rotors properly. With this model of hub, assembly is very easy.

As purchased, the hub has four studs pressed into it to accept the lug nuts that hold a trailer wheel rim on. The first step in preparing the hub is to knock those out with a hammer. It goes quite easily—chuck the hub up in a sturdy vise, and simply pound them out!

Hardware list

(All hardware is stainless steel except as noted)

- 1/2 inch – 13 tpi nuts, quantity 25
- 1/2 inch – 13 tpi acorn nuts, quantity 7
- 1/2 inch – 13 tpi threaded rod 6-7/8 inches long, quantity 4
- 1/2 inch – 13 tpi threaded rod 4-1/4 inches long, quantity 3
- 1/2 inch lock washer, quantity 18
- 1/2 inch washer, quantity 6 (We use the smaller AN style. If the air gap must be shimmed out wider, 6 more of these may be required.)
- 1/2 inch – 13 tpi threaded rod 10 inches long, quantity 4. Do not need to be stainless.
- 1/2 inch – 13 tpi nuts, quantity 8. Do not need to be stainless.

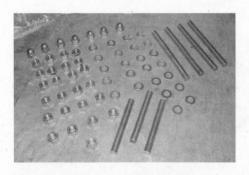

Figure 14.6—All the hardware needed for assembling the alternator.

Figure 14.6 at left shows all the hardware required to assemble the machine except for the bolts and nuts that hold the tail on. We prefer to use all stainless steel hardware for better quality threading and so that it can never rust, but this is optional except for the stator mounting hardware near the edge of the magnet rotors—these parts *must* be stainless or else their magnetic attraction to the magnet rotors will put drag on the machine. Nuts and lock washers are used as spacers between the magnet rotors, so it is important to use fairly high grade hardware and make certain that all of these spacer nuts and washers are the same thickness. Cheap nuts and washers may vary in thickness, but stainless and SAE hardware is usually very consistent. The acorn nuts are not essential—you could use regular nuts, the acorn nuts simply look better.

One of the steel hubs that "sandwich" the blades between them is shown in Figure 14.7 at right. You've already built it from the *Frame* chapter. You'll be using it as a jig for assembly, otherwise it will not be used until the blades are installed.

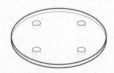

Figure 14.7—Front rotor hub.

The magnet rotors must be cleaned before assembly. Oftentimes while building them, or simply in storage around the shop, bits and chips of metal from grinding and drilling get stuck to the magnets. The chips lay flat when the rotor is sitting on the workbench, but after the alternator is assembled all the chips will stand up and rub on the stator. A good way to remove metal chips from the magnet rotors is with duct tape, as demonstrated by George in Figure 14.8 at right.

Figure 14.8—Cleaning metal bits off the magnet rotors with duct tape.

Figure 14.9—Studs.

You are now ready to start assembly. Take the four pieces of 6-5/8 inch long threaded rod and put one nut on each. Screw the nut down on one end so that there is 7/8 inch of threaded rod behind it, as shown in Figure 14.9 at left. Next, insert the long end of the stud through the backside of the wheel hub. Put on a lock washer and another nut, and tighten the stud to the hub—no need to get things tight here, just finger tight is what you want. Do this with all four studs, and then run a nut onto the front of each stud so that about 1/4 inch of threaded rod is protruding.

Figure 14.10 at right shows what things should look like now. Looking at one stud and starting at the front (going left to right in the picture), here is what's important: 1/4 inch of threaded rod, one nut, more threaded rod, another nut, a lock washer, the wheel hub, a nut, and then 7/8 inch of threaded rod protruding.

Take the back magnet rotor (the one with only the four stud holes) and turn it so that the magnets face down on the bench. Be sure the bench is clean of any metal bits. Don't do this on a steel workbench!

Figure 14.10—Studs inserted into the trailer hub assembly.

Now take the back of the hub and put it into the center hole of the magnet rotor, so that the ends of the studs (the ends that are 7/8 inch long) are poking into the four holes. This helps to align things before tightening anything. Take one of the steel rotor hubs (the one shown in Figure 14.7, previous

Figure 14.11—Aligning and tightening the stud and hub assembly.

page) and place it over the studs on the top. It should come to rest on the four nuts—that's why those nuts are there with 1/4 inch of threaded rod sticking out. Using the steel blade hub for alignment assures that the studs are well aligned top and bottom and that everything will fit together nicely, as shown by George in Figure 14.11 at left. Now you can tighten the nuts on both sides of the wheel hub. You should get them very tight, and the goal is to move only the top nut (the one with the lock washer under it) and not the bottom one, so that you are sure to have 7/8 inch of threaded rod exposed behind the back nut.

Remove the steel blade hub disc (used only for alignment for now) and the four nuts that hold it there at the top, and then remove the entire hub assembly from the magnet rotor. Turn the hub upside down (so the back is facing up and there is 7/8 inch of threaded rod sticking up). If the studs were cut accurately and the nuts positioned properly, the hub should sit pretty flat on the ends of the studs. Carefully lower

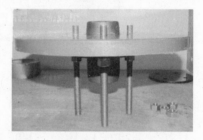

Figure 14.12—Assembly turned upside down on the workbench.

the back magnet rotor down onto the hub so that the studs go through the holes. The magnets should be facing down, and it should now look like Figure 14.12 at right.

Figure 14.13—Tighterning the nuts on the back of the rotor.

Remember that the hub is steel and the magnets will attract it powerfully! If you bring it down right on center it will be fine, but if things are off center the rotor will grab the hub and you'll have to try again. Sometimes it's handy to have someone holding the hub down for you to keep things positioned. On the back side of the back magnet rotor each stud gets one lock washer and one nut, as shown in Figure 14.13 at left. Tighten these just a bit with a wrench now—after the alternator is together you'll tighten them completely.

Grease the bearings with axle grease before assembly. Press lots of grease into both bearings, and work the rollers around for a while and be sure everything is covered. Grease the spindle up well also. Use lots of grease, as demonstrated by George in Figure 14.14 and Figure 14.15. You won't be able to add more grease until a year from now, when you take your turbine down for its first yearly maintenance—so pack it full!

Figure 14.14—Greasing the bearings.

The front and back bearings are identical for this wheel hub. Put one of the bearings on the wheel spindle and push it back all the way. Usually when you order a hub, it comes with a seal. We do not use the seal in this wind turbine, it creates too much friction and prevents easy startup. If you grease the bearings well they will be fine for quite a long time, and during each scheduled yearly wind turbine maintenance session you should re-pack them with grease.

Figure 14.15—Greasing the spindle.

Carefully pick up the hub/back magnet rotor assembly by the studs and place it onto the spindle up against the back bearing, as shown in Figure 14.16. Then insert the front bearing. Press a little more grease into the front of the bearing—it can't hurt! Place the washer in front of the bearing, and then tighten the nut over the bearing, as shown next page in Figure 14.17. Get the nut reasonably tight, and then back it off so that you can insert the cotter pin. Once the cotter pin is in, back the nut off as much as the cotter pin will permit. You don't want the nut tight, there should be a touch of play. If you test this, you will feel extra drag when the nut is

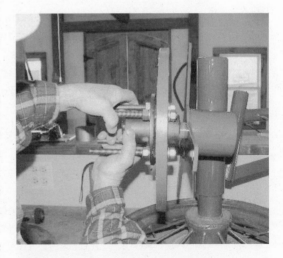

Figure 14.16—Placing the back rotor on the spindle.

tight, and a more free-spinning rotor when it is slightly loose. Slightly loose is what you want, because the machine will not start up easily and the bearing will wear out prematurely if it is too tight.

Figure 14.17—Back magnet rotor, front nut and cotter pin in place.

This is a good time to spin the hub and check to see that the magnet rotor runs reasonably true. Magnet rotors cut from 1/4 inch steel are rarely perfectly flat, so you can expect a bit of wobble. This won't hurt anything, but the less the better. If there's something seriously wrong, now is the time to catch and correct it!

Next up is mounting the stator to the machine. Take the three 4-inch long pieces of threaded stainless steel rod and screw an acorn nut down all the way on one end of each, then put one of the stainless washers on. Insert the studs through the three 1/2 inch holes in the stator. On the back side of the stator install another stainless washer, a lock washer, and a nut. Finger tighten these, you don't want them real tight yet. Then run one more nut down about 1 inch on each threaded rod. The result is shown in Figure 14.18. Fit the assembly onto the wind turbine, Figure 14.19. The three studs should fit through the three holes in the stator bracket. You can now adjust the back nuts (the ones against the stator bracket) with your fingers and set an approximate air gap (about 1/8 inch) between the stator and the back magnet rotor.

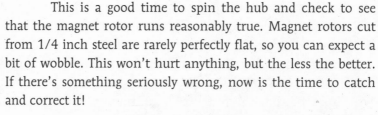

Figure 14.18—Hardware attached to the stator.

Figure 14.19—Stator mounted on frame.

Now tighten the nuts on the stator with two wrenches as shown in Figure 14.20. Get these reasonably tight, but not so tight that they crack and ruin the cast resin stator. You won't touch these nuts again—any further adjustments will be made at the stator bracket, not the stator.

Figure 14.20—Tightening the stator mount nuts.

Where the studs come out through the back of the stator bracket, install a lock washer and a nut. Using two wrenches (shown in Figure 14.21 at right) you will set the clearance between the stator and the back magnet rotor. About 3/32 inch is good at this stage of assembly.

Figure 14.21—Tightening stator mounting studs to the frame, which sets the clearance to the back magnet rotor.

Next, check and make sure that the stator is well-centered inside around the wheel hub. You don't want it rubbing here. If it's

Figure 14.22—A nicely centered and mounted stator.

not well-centered and you are sure the holes were drilled correctly in the stator, then the problem is usually because either the nuts are not tight yet, or perhaps the stator bracket is warped. If needed, you could bend the stator bracket a bit. If the assembly is not centered well, it should be simple to find the problem and adjust for it somehow. It's very important that the stator and magnet rotor have good clearance here. Figure 14.22 at left shows a nicely-centered stator, ready for the installation of the front magnet rotor.

Jacking screws

The four jacking screws you will need to make are pictured in Figure 14.23 at right. These are tools, not part of the wind turbine. Cut them about 10 inches long. They are made from 1/2 inch – 13 tpi threaded rod and don't have to be stainless. Double-nut (jam two nuts together tightly) each on one end. The other end needs to have a slight chamfer ground on it—this keeps them from getting "mushroomed out" and difficult to extract when they squish against the wheel hub. After cutting the jacking screws to length, grind off the burrs and clean up the threads with a file. Test them by running a nut up on the end which was chamfered, as shown on the rightmost jacking screw in Figure 14.23.

Figure 14.23—The jacking screws.

Run the jacking screws into the front rotor so that about 3 inches sticks out on the magnet side, as shown at left in Figure 14.24. Try to keep them very even (all of them poking out the same amount) or else things will bind up as the rotor goes on. If paint is clogging up the jacking screw holes in the front magnet rotor, clean the threads with a tap. A bit of oil on the screws doesn't hurt either, because assembly is nice and peaceful when these jacking screws turn easily by hand. Remember your index mark, punched into each magnet rotor when you were building it? Turn the back magnet rotor so that the index mark is pointing straight up (12 o'clock noon) and make sure the front magnet rotor is aligned the same way, so that its alignment mark is also in the noon position.

Figure 14.24—Jacking screws in place.

Carefully pick up the front magnet rotor by the jacking screws (as shown in Figure 14.25 at left) and place it over the studs sticking out of the alternator. Do not get your fingers near the magnets, or between the rotor and the stator! Ever! The jacking screws make a nice safe handle. Push it onto the studs and you'll feel the magnets "grab" each other with magnetic attraction. It should pull down so the four jacking screws come to rest on the hub.

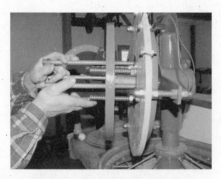

Figure 14.25—Carefully lowering the front magnet rotor onto the studs. The jacking screws stop it.

Now, use a wrench and a bit of patience to lower the front rotor down (Figure 14.26 at left) by alternately turning the four jacking screws, trying to keep them even. Usually it works best to go around in a circle, giving each jacking screw about one turn and then moving to the next, until the front rotor comes to rest on the four nuts which hold the studs to the wheel hub. This Dexter hub model and this arrangement of nuts and washers as spacers makes for a nearly perfect air gap if you cast your stator to exactly 1/2 inch thick. If the stator is thicker than that, you may need to remove the front rotor and put washers behind it as shims, though in most cases that's not necessary.

Figure 14.26—Safely lowering the front rotor down with the jacking screws.

At this point the alternator is assembled! It's shown in Figure 14.27. The clearance between each rotor and the stator should be around 3/32 inch. A clearance of as little as 1/16 inch is probably acceptable, but any less and it would be wise to take it apart and add washers as shims to widen the gap. Once both rotors are on, you can fine tune the position of the stator using the nuts on both sides of the stator bracket. At this point, all adjustments are done by moving the stator, not the magnet rotors. Now you can do some initial tests on your alternator, to make sure everything was wired, cast and assembled correctly. The alternator should spin freely with little wobble, and the magnet rotors should not rub at all on the stator.

Figure 14.27—Completed alternator, with alligator clip leads attached for testing.

Testing

Your first question now will undoubtably be, "does it work?" There are some simple tests you can do right away to make sure everything came out right. A good first electrical test is to short one power output terminal to another with a jumper wire and spin the alternator by hand. When any phase is shorted to any other, it should feel stiff, but kind of lumpy to turn (it will "cog"). When all three phases are shorted together it should become very stiff and difficult to turn, yet very smooth, with no "bumps." If you don't feel this—it's time to start checking for errors!

If you don't feel any physical resistance to turning the alternator when any phase is shorted to another, first check to be sure your stator output wires are well-connected to their terminal lugs—you are not making any power and might have a loose connection inside the stator. If you feel lots of resistance when the output terminals are *not* jumpered together, you might have an internal short in the stator. If it feels steady, all of the phases are shorted, if "lumpy" it means a single phase is shorted to another. It's sometimes possible to fix internal shorts or loose connections if they are located near the inner rim of the stator—you can do some creative "dentistry" with a die grinder, expose the problem area, and possibly fix the problem by re-soldering a loose connection or re-insulating a short. Then you can fill up the cavities with epoxy to seal them up. If this exploratory surgery fails, your only option is to start winding more coils and build another stator.

Another quick test you can perform on your new alternator is to simply connect a low-voltage 12 volt incandescent light bulb up to two of the output terminals. As you spin the alternator by hand, the bulb should flicker as AC current passes through it. If you spin too fast, you'll burn out the bulb!

The most important test you can perform is for cut-in speed, which is crucial to the operation of the machine. It's defined in this sense as the rpm at which the alternator's output voltage equals your system's battery bank voltage and starts charging. We are shooting for a cut-in speed of 140 rpm with this design. If you made some errors or sloppy alignment during construction, your cut-in speed might be a bit higher, and if you worked with surgical precision it might be a bit lower—as long as it's within 20 rpm of the goal of 140 rpm, you'll be fine.

To accurately test for cut-in speed you'll need a multimeter. If you are unfamiliar with how to use a multimeter, the sidebar here will give you some pointers. It's also handy to have an laser tachometer, though you can get very rough results by estimating with a stopwatch how fast you are spinning the alternator. You'll also need to get out the rectifier assembly you built from the previous chapter. And, it helps to have an assistant to spin the machine. Connect each of the three alternator output terminals to the AC input terminals on the rectifier assembly. Connect your multimeter, set for measuring DC voltage, to the DC output terminals. Now, slowly and steadily start turning the alternator. When the output voltage reaches your battery bank voltage, call out the rpm from the tachometer. That's your cut-in speed!

How to use a multimeter without letting the smoke out

As we explained in Chapter 2 (*RE 101*), electricity consists of "magic smoke" that is placed inside tiny black chips at a big factory. If you make a mistake and let the smoke out, it's impossible to put it back in again at home—you need special equipment for that. Seriously, though, you need to own a multimeter if you have a renewable energy system! And it's very easy to accidentally let the magic smoke out of any multimeter, no matter how expensive. Plus, we've observed that many people who own a multimeter don't really know how to use it correctly.

Ranges

Take a look at the dial in the typical multimeter pictured in Figure 14.28. The one shown will measure ohms, volts, and amperes. Each section of the dial also gives different "ranges" that the meter can measure in. The number given is the *maximum* measurement it can make when set there. For example, if you set it in the 2 volt range and connect it to a 12 volt battery, the meter will show something like the "-1" like in the picture—the voltage is out of range. Turn the knob to the 20V range instead and you'll get an accurate reading. There are also "auto-ranging" meters out there with which you don't have to worry about this, they set the range automatically for you. Setting the meter in the wrong range won't usually damage it—but the cheaper the meter, the less internal protection from this sort of event it will have inside. If you have it set for a range that's too high—like set in the 200 volt range while measuring a 1.5 volt battery—the reading will be very inaccurate since you'll lose decimal places. For example, such a measurement might show 2 volts while in the 200 volt range, 1.6 volts when you switch to the 20 volt range, and finally show a very accurate 1.554 volts when set to the 2 volt range. That's why you want to set the range for just above what the measurement will be! Also note the "AC-DC" switch—this also must be at the correct setting or the reading will be wrong.

A meter won't show you the exact measurement at any given instant, especially with AC circuits where the measurements are changing at 60 times per second or more. So the meter has to average things out, otherwise the display would be changing so fast you couldn't follow it. There are different

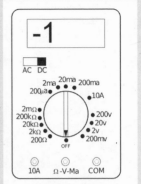

Figure 14.28—A typical digital multimeter.

ways for a meter to do this. Most folks with RE systems use a "true RMS" meter (RMS stands for "root–mean–square") because it will give more accurate readings from the sometimes odd waveforms produced by inverters and wind turbines. See the *Meter madness* sidebar in Chapter 20, *Scaling it up and down*, for more metering information. True RMS meters cost a bit more, but are worth it in RE applications.

Really inexpensive multimeters can vary significantly in even their simplest readings; it's better to invest money in a good one. Cheap meters may drift significantly if their internal battery gets low, too. Better quality meters won't keep trying to work with inadequate internal power—they'll instead shut down and tell you to replace the batteries before continuing.

Jacks and fuses

You may also have to change which jacks your meter's test leads are plugged into. The black lead always plugs into the jack marked "Common" or "COM." The red lead will plug into the jack marked "V-ohm-ma" for measuring volts, ohms or milliamps, and into the jack labelled "10A" for high amperage measurements (10A meaning a maximum of 10 amps). The more expensive your multimeter, the more likely that various internal components are protected by fuses, which (hopefully) will blow before the meter is damaged. Cheap meters will release their magic smoke at the slightest breach of measurement etiquette. The fuses are cheap to replace, but they won't be standard sizes you could find at the local auto parts store—you may have to special order them or try Radio Shack or another local electronics store.

Measuring voltage

To measure voltage, first make sure the multimeter is set correctly for AC or DC—it's usually a switch on the front plate. It might be labeled "~" and "---" instead of AC and DC. Make sure the meter is set for close to the proper voltage range, and make sure the black lead is plugged into the "common" jack and the red lead into the "V-ohm-ma" jack. Then simply connect the test leads to the circuit you want to measure. Remember that the "open-circuit" voltage of a power source like a wind turbine or solar panel will be a very different measurement from the "closed-circuit" voltage when the unit is connected to a battery bank! To get open-circuit voltage, you'll have to completely disconnect the turbine or PV panel from the battery bank. Figure 14.29 shows a closed-circuit voltage measurement, with a typical reading. Connected like this, you are really reading the battery bank's voltage, since the batteries are "clamping" the output down to their own level. Figure 14.30 at right shows how you would connect the meter to measure just the solar panel or wind turbine alternator open-circuit voltage. Note that the meter is hooked up with backwards polarity in the

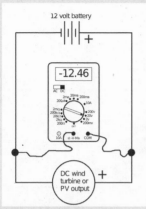

Figure 14.29—Measuring closed-circuit voltage.

drawings. This doesn't hurt anything, but it's why the meters are reading minus 12.46 and 39.7 volts instead of plus.

Measuring resistance

First, make sure the circuit you want to measure is *not* energized—for resistance measurement, the multimeter uses its own internal power source, and it could be damaged if connected to a live circuit when set in the resistance ranges. Also make sure the component you want to measure is isolated from the rest of the circuit—for example a resistor or single coil will need to be disconnected on at least one side or else your

Figure 14.30—Measuring open-circuit voltage.

measurement will be of the entire circuit, not just the component you are interested in. Set the range close to what you predict the measurement will be and connect the meter. If the circuit is live, smoke will come out of your meter to indicate that you are not supposed to do this!

Measuring amperes

It's important when measuring amps to have some idea of what your expected reading will be. If in doubt, plan on high amperage. The jack labelled "V-ohm-200ma" in Figure 14.28 is what you'll use for measuring low amperage, and the "10A" or 10 amp jack is for anything over that 200ma mark. If you accidentally try to measure current over 200ma (or whatever your meter's milliamp range is) with the wrong jack, you'll fry the meter! Between the two co-authors of this book, we have a pile of fried multimeters that numbers over a dozen—and we always *try* to be careful. If there's a possibility of current over the maximum milliamp range of your meter, be *sure* to connect it first via the high-current jack instead. If the current is under 200ma, you can then safely change to the "V-ohm-200ma" jack.

To measure amps, you'll need to connect the meter in-line between one of the circuit's outputs and a load as shown in Figure 14.31. Usually the load will be a battery or battery bank. Note that this means discon-

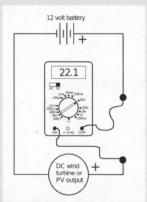

Figure 14.31—Measuring amperage.

necting wires in the system so your test leads will be in the circuit. If you connect the multimeter directly across the outputs, you are measuring short circuit current. This may be what you want (for example the maximum output current of a solar panel), but with a wind turbine it's most likely not what you care about. When testing renewable energy equipment, keep in mind that the current flowing into an actual battery bank is usually what you are interested in.

When measuring amperes, it can be very inconvenient to disconnect and re-connect wires to put the meter in line in the circuit. Clamp-on multimeters (or adding a clamp-on probe to a normal multimeter) are very handy for this! Cheaper ones will measure only AC current with the clamp, better models will do both AC and DC.

15. Blades

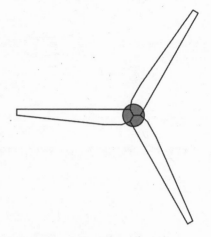

The blades are an important part of any wind turbine—they are the "engine" that drives the alternator, "fueled" by moving air. There are entire books written about wind turbine blade design, and if you are interested in the intricacies and math involved, check our *Sources* chapter for some book recommendations. Blade design, however, boils down to just a few important factors:

- Maximize lift, and minimize drag.
- Match the swept area to how much power the alternator can produce.
- Match the tip speed ratio to the rpm at which the alternator is most efficient.

The wind turbine blades you'll be carving here are a compromise between the factors above and factors of strength, cost, availability of materials and ease of construction. The blades are sized, tapered and twisted to give a tip speed ratio (see Chapter 3, *Power in the wind*) of around 6, where this kind of turbine runs most efficiently (Chapter 3, Figure 3.11). They sweep the correct amount of area for the power capabilities of the alternator you've built. And, they use a simple and easy-to-carve airfoil that's practical to produce in the home workshop with hand or portable power tools.

Blade terms

Before we start, let's define a few terms. Figure 15.1 shows a blade from the front, or "face"—the axis the wind comes from. Figure 15.2 shows a blade from the tip or root, what we'll call the end view or tip view. The "tip" of

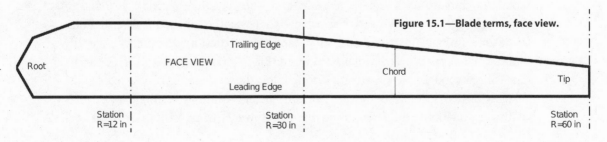

Figure 15.1—Blade terms, face view.

Root · FACE VIEW · Trailing Edge · Leading Edge · Chord · Tip

Station R=12 in · Station R=30 in · Station R=60 in

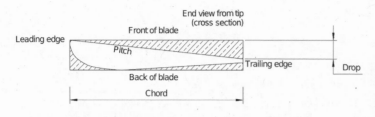

Figure 15.2—Blade terms, tip view (cross section).

the blade is the end that's at the very outer diameter—farthest away from the alternator. The "front" of the blade is the surface that faces towards the wind, it's flat and angled a bit. The "back" of the blade is facing away from the wind and it's rounded into a curved airfoil shape. The "root" is the inside of the blade, closest to the hub and the alternator. The "leading edge" is the front edge of the blade in the direction it's spinning. It gets to the destination (wherever that is) first—if it were an airplane wing, the leading edge would be the front edge of the wing. The "trailing edge" is the edge of the blade that gets there last—if it were an airplane wing it would be the back edge of the wing.

The "pitch" of the blade is the angle between the surface of the front of the blade and the plane of the blade's rotation. It changes over the length of the blade, giving the blade a twist—lots of twist near the root, and very little near the tip. The "chord" of the blade is the width (the distance between the leading edge and the trailing edge) and it's wider at the root than at the tip. The "thickness" of the blade is measured at the fattest point in the airfoil. A "station" is one of the points you will measure out to from the root when laying out the cuts to be made in the wood. The "drop" is a measurement of how much material to remove by carving, made when laying out the blade for carving the pitch into the front.

Blade materials

We make wooden blades here—wood is the original, cheapest and most common "carbon-fiber composite" available! Some enthusiasts use exotic foam cores, layered up with expensive resin and fiberglass like a small airplane or boat hull. You can make very nice blades that way, but it's expensive, tedious, time-consuming and toxic—and such exotic construction doesn't provide any substantial advantages over plain old wood. Blade strength is not much of an issue since wood is plenty strong enough for the stresses encountered by the blades during normal operation. The only blade failures you might encounter with this 10-foot turbine would be due to a tower, furling system or alternator flaw that would allow the spinning blades to contact the tower or guy wires, and in that case it won't matter what materials you built them out of—they will disintegrate. Metal blades are not recommended as they are prone to fatigue cracking and can turn into heavy, high-speed projectiles if something fails.

The most common material to build blades from is common pine or fir "2×8" boards from a lumberyard. Typically a planed 2×8 is actually 7-1/2 inches wide and 1-1/2 inches thick, so the plan is based upon those dimensions. Pine and fir have good strength to weight characteristics. Very hard and heavy or very soft and light woods should be avoided. The very best choice is probably clear (knot-free) Sitka spruce, but it's expensive and hard to find. Most lumber yards have perfectly acceptable material. Generally you should use conifers, although the very lightest ones might be too weak. Avoid redwood if possible. The wood should be dry and as knot free as possible. We often build our blades from Western red cedar 2×4s that we laminate together with urethane glue, but that adds an extra step that is not necessary—it does make for a strong, lightweight blade though. Whatever you find, you need three boards about 7-1/2 inches wide, 1-1/2 inches thick and 60 inches (5 feet) long.

Tools

There are a variety of tools you can use for carving blades. A quality draw knife is essential (Figure 15.3). Avoid cheap drawknives, any money you save will come right back to bite you as utter frustration! Chisels, hammers, sandpaper, planes and other wood

Figure 15.3—A quality draw knife is essential for blade carving. This one is made by Two Cherries and costs about US$80.
Photo courtesy of Wilh. Schmitt & Company.

working and carving tools are handy. Some of the work at the beginning of the project involves removing large pieces of the board, and a band saw is very useful, but not essential. A handheld power planer saves lots of time, but hand planes work almost as well and they're much quieter and more peaceful to work with. A few hours of living with the loud, high-pitched whine of a power planer can make even the cheeriest person or dog grouchy and irritable! The whole blade carving process can be done fairly easily with hand tools only.

Figure 15.4—A hazard of blade carving.

The instructions below will describe how to carve a single blade. You'll need to make three of them. We suggest you make all three at once rather than making one blade at a time. There are several procedures to be performed on each blade, and it's better to do only one operation to each blade and work them all along together, they'll come out more alike that way. In other words, if you perform an operation on one blade—do it to the other two before you move along to the next step.

Find some lumber

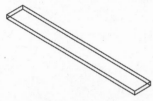

Start with three 2×8 boards, 7-1/2 inches wide, 1-1/2 inches thick and 60 inches (5 feet) long (Figure 15.5). Hopefully they are free (or mostly free) of knots. Try to pick lumber that's not warped, with nice straight grain—the more vertical the grain the better.

Figure 15.5—Your basic 2×8 board.

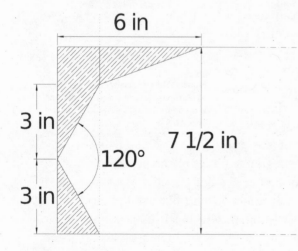

Cut out the blade root

Figure 15.6 shows the root of a blade. This drawing will help you with the layout. You need a 120 degree angle at the root so that all three blades fit together tightly. Lay this out on all three blades and cut the root out. You can use a band saw, circular saw, or hand saw—just make sure the cuts are very precise.

Figure 15.6—Laying out the cuts for the blade roots.

Cut out the shape of the blades

Figure 15.7 shows the shape of the blades. You can see how the chord of the blade is tapered, widest at the root and narrowest at the tip. At the tip (radius(R)=60 inches) it's 3 inches wide. At the halfway point (R=30 inches) it's 6 inches wide. Draw a line between those two points and extend it to where it meets the edge of the board—this will be somewhere around R=14 inches, but it can vary slightly depending on the width of your lumber. You can either make a template and trace it onto all three blades, or just lay it out on one blade and cut out the profile, then trace it to the other three from the first blade. Cut out the blade shapes with a band saw, circular saw or whatever you happen to have available.

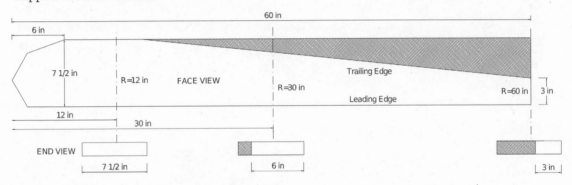

Figure 15.7—Layout for cutting out the shape of the blades.

Taper the thickness of the blade

As the blade gets narrower towards the tip it also gets thinner. Figure 15.10 shows the layout for tapering the thickness of the board. The top of the image is the front of the blade and you won't touch that part at all during this step—all the material is removed from the bottom of the blade. The edge view is looking at the board from the leading edge, the end view shows the cross section of the blade (white) and the scrap (shaded) at R=60 inches (the tip), R=30 inches and R=12 inches. It's better to be too thick than too thin at this point—be sure not to get things too thin or the blade will be weakened. The dimensions we give for thickness in the drawing are the absolute minimum.

Figure 15.8—Dave M. tapering the thickness of a blade using a band saw.

A band saw is probably the best tool for cutting the board thickness, and Dave is demonstrating how to do it in Figure 15.8. Give yourself room for slop—don't crowd the line. When cutting this with a band saw it is possible that the board will not be perfectly square with the table, and it's easy to end up with a different thickness depending on which side is down on the saw table, so give yourself room! After sawing, it's best to finish the job with a hand plane or a power planer to get the thickness of each blade down to the exact dimensions needed.

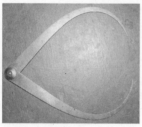

Calipers (Figure 15.9) work well for making sure you've got the thickness right throughout the length and width of the blade. If you don't have a band saw the whole job can be done with a plane or planer (or even a draw knife)—it just takes longer and makes more of a mess on the floor. When finished the back of the blade should be smooth and square with the sides.

Figure 15.9—Calipers

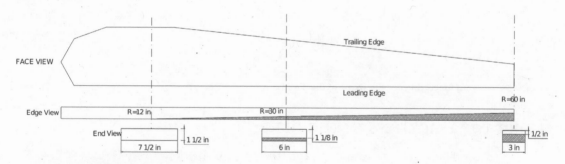

Figure 15.10—Layout for tapering the thickness of the blades.

Carve the pitch on the front of the blade

At the tip of the blade the pitch is about 3 degrees. At the center (R=30 inches) it's about 6 degrees. Near the root, where the carving stops, it's as steep as the board will allow. The drawings do not show the angle of pitch—instead they show the amount of wood that must be removed from the blade along the trailing edge, (the "drop"), because that's what you'll use to lay out your cuts at the trailing edge. You won't carve any wood at all out of the leading edge at this stage.

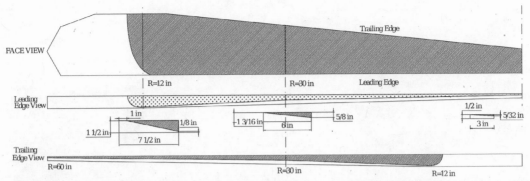

Figure 15.11—Layout for carving the pitch into the front (face) of a blade. In the leading edge view, you are looking *through* the wood of the leading edge to the cutout behind. In the trailing edge view, the blade has been reversed and the cutout is facing you.

Figure 15.11 (above) that shows all this might seem a bit confusing. In the image (center, edge view) you're looking at the leading edge of the blade, and the darkened area is the material that must be removed from the trailing edge. Turn your unfinished blade around so you're looking at the trailing edge, measure down from the front of blade, and make marks at R=12 inches, R=30 inches and at the tip (R=60 inches). At R=12 inches you'll be removing all but 1/8 inch of the wood from the board—in other words the pitch will be so steep here that it occupies almost the full board thickness, and you'll leave that 1/8 inch only for a bit of strength. At R=30 inches you need to measure down from the top of the board 5/8 inch (the drop). At the tip measure down 5/32 inch and make your mark there. Then connect the dots and you'll have a line along the trailing edge of the blade to carve down to.

To carve the pitch of the blade you'll be carving between two lines. One line is the one you just drew along the trailing edge, the other is the leading edge of the blade (the corner between the front of the blade and the leading edge). Do not disturb that corner of the board, but use it as a line. The edge view drawings in Figure 15.11 show this.

Things get a little different as you approach the root. In the drawing, the carving stops near the root of the blade, and you need to draw a line here past which you'll not be carving anymore. The exact shape and location of this is not critical, but you need to make all three blades the same. It's best to make a template for this and trace it onto all three blades. As the carved surface approaches this curved line (near the root) it will taper up to full board thickness. Near the root, where the carving is deep and it tapers out into the thickness of the board, it's sometimes best to cut lots of slots (kerfs) between the lines with a hand saw (one every inch or so) and then remove the wood between the slots with a chisel and mallet (Figure 15.12, next page).

Figure 15.12—Roughing out the root area with saw kerfs, a chisel, and a mallet.

Once you get started it should all seem fairly easy. A draw knife (Figure 15.13) is a great tool for removing lots of wood fast. A hand plane or a power planer also make pretty quick work of things.

Figure 15.13—Blade carving with a draw knife.

One you've carved down right to the line, smooth and flatten the surface with a plane and/or sandpaper. Check with a straightedge between the leading and trailing edges—the blade surface should be flat.

Carving the backs of the blades

All that's left to finish the job is to carve the airfoil profile on the back side of the blades. As a rule the thickest part of the airfoil is always one third of the way back from the leading edge. In other words, the blade is 3 inches wide at the tip, so the fattest part of the airfoil will be 1 inch back from the leading edge. For example, at R=30 inches the blade is 6 inches wide, so the fattest part of the airfoil will be 2 inches back from the leading edge. Another general rule is that the airfoil is about 1/8 as thick as it is wide. This ratio changes gradually as you approach the root, so that near the root it's about 1/6 as thick as it is wide. This is not terribly critical, so long as you're close, but you never want to be less than 1/8 as thick as you are wide. At the tip the blade is 3 inches wide, so the thickest part of the airfoil should be about 3/8 inch thick. Anything between about 3/8 inch and 1/2 inch will be OK at the tip, it should not be more or less than that though.

A draw knife is a good tool for roughing out the back side of the blade. This can also move along very quickly with a hand plane or a power planer. So, turn the blade over so that you're looking at the back. At R=12 inches, measure back 1/3 of the way from the leading edge to the trailing edge (the blade is 7-1/2 inches wide at this point, so measure back 2-1/2 inches from the leading edge) and mark it. Do the same thing at the tip (it's 3 inches wide so measure back 1 inch). Draw a straight line between the marks. This line marks the thickest part of the airfoil and this should never be disturbed by carving. This is the one place on the back of the blade where you'll do no carving! In Figure 15.16, Matt is shaping his blade backs with a power planer, and in Figure 15.15 two of our seminar students are doing it with draw knives.

Between this line and the leading edge you need to carve a nice rounded surface as shown in the end views from Figure 15.14. Between the line and the trailing edge it could be a slightly rounded but almost a flat surface. As with all other steps, it's best to do one operation to each blade so they come out the same. We find that even a change in mood can affect your final work, so it's good to work along all three blades at the same time! Once you have this roughed out, then finish it with sandpaper. The leading edge of the blade should be rounded. There should be no sharp edges between the front of the blade and the back of the airfoil. As the airfoil approaches the root of the blade (around R=12 inches), it should just taper out to the original profile of the board. This area involves some concave surfaces which are impossible to do with a plane. Your best bet is to use a draw knife or a spoke shave. The trailing edge should be brought down fairly thin (about 1/16 inch) and fairly sharp. Ideally it should be very sharp, but you don't want to make it so thin that it's too fragile—1/16 inch thick is about right for the trailing edge.

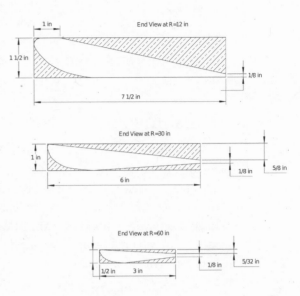

Figure 15.14—Layout for carving the airfoil in the blade backs.

Figure 15.15—Carving blade backs with draw knives.

Figuer 15.16—Carving a blade back with a power planer.

If you ran into any knots in the wood, you might have holes in the blades. You can fill these with wood putty, then sand it down so it's even again. Finally, sand all surfaces of the blades so they are smooth, down to 220 grit.

Finishing

There are many ways to finish your blades after you are done. Our method is simple and cheap—we use multiple thick coats of boiled linseed oil. This has worked fine in our mild Colorado climate, and since we always drop our wind turbines once a year for maintenance we can re-apply more coats of linseed oil every year. If you are in a climate with lots of flying dust, snow or salt spray, you may want to consider a more durable finish. We'll talk more about applying the linseed oil in the next chapter, *Rotor assembly*.

Some commercial wind turbines use a special kind of plastic tape on the leading edge to protect the blades from dust and snow erosion. The product was originally developed by 3M for protecting helicopter blades. It's very expensive, and the manufacturer's instructions must be followed exactly or it will do more harm than good if it later starts to peel off. Check our *Sources* chapter for information on where to find leading edge tape if you are interested.

Most commercial wind turbine manufacturers either use plastic blades that are unpainted, or paint their wooden blades. If you choose to paint your blades (to protect them from an extreme salt air climate, for example), you'll need to use the best and most expensive automotive primer and enamel you can find, and follow the manufacturer's instructions exactly! With painted blades, any cracks or gaps in the paint job will allow moisture in and trap it next to the wood. This again will do more harm to your blades than good, and you should check the paint carefully for such issues every year when you lower the turbine for maintenance.

DOG HAIKU
Engineers design,
but they rarely carve the blades.
Dogs and cats, they nap

16. Rotor assembly

At this point you've already carved three nearly identical wind turbine blades (Figure 16.1). This chapter will detail how to assemble them into a rotor. Your wind turbine is almost complete! Following is the list of materials and hardware required to assemble the blades. Once again, there are many stages of this process that could be done differently, we're just showing how *we* do it. Some folks prefer to glue the blades and hubs together—however, we recommend using screws as shown here so that you can disassemble the blade set should you ever need to replace parts.

Figure 16.1—Three blades, ready to be assembled into a rotor.

Rotor assembly parts list

- 1-1/4 inch galvanized wood screws, quantity 60
- 5/16 inch – 18 tpi, 3 inch long carriage bolts, quantity 6
- 5/16 inch washers, quantity 6
- 5/16 inch lock washers, quantity 6
- 5/16 inch – 13 tpi nuts, quantity 6
- 11 inch diameter disk of 1/2 inch plywood (preferably Baltic birch), quantity 2

The wind turbine blades will be sandwiched between two plywood hubs, 11 inches in diameter (Figure 16.2). Baltic birch plywood is much stronger than normal plywood and holds up to weather well—we prefer to use it for the blade hubs. Normal plywood will do if Baltic birch is difficult to find in your area. It is only rarely available from normal lumber yards, but you can find it at shops that sell more exotic hardwoods for cabinet making and fine woodworking. It's not terribly expensive.

Figure 16.2—Blade hubs, made of Baltic birch plywood.

Divide the hubs into three parts and use a compass to neatly lay out where the screws will be located before you drill the holes. Drill about 15 holes in each area where the roots of the blades will be between them, as shown in Figure 16.2. It's easiest to drill both hubs together. Then countersink the holes so the screw heads will be flush with the top of the plywood. Drill a small hole (about 3/16 inch is good) in the center of both hubs—this will make centering and aligning the two hubs easy.

Figure 16.3—Blades laid out on the floor for alignment.

Lay out the three blades on the floor with the flat sides (the sides that will face into the wind) up, as shown in Figure 16.3. If you cut the 120 degree angles accurately when fabricating the blades, they should fit together fairly tightly. There should be a tiny flat spot on the center of each blade root where they meet each other. This is so that when they come together there is a hole in the center of it all big enough for a small drill bit as an alignment pin, to help make sure you get everything (the blades and hubs) centered perfectly.

Center one of the birch hubs over the blades and line up the screw holes you drilled over the blades (Figure 16.4). This is where you'll use the small drill bit mentioned above in the center hole. It's helpful if this hole extends into your floor, too—we have a hole drilled in the middle of our shop floor just for this purpose!

Figure 16.4—Centering the first hub.

Figure 16.5—Initial attachment of hub with one screw.

Put one screw through the hub into each blade (Figure 16.5). This will hold it all together just tightly enough to allow you to make more precise adjustments in alignment.

Use a tape measure and measure the distance from tip to tip between the three blades, as demonstrated by George and Rich in Figure 16.6. You'll be able to make small adjustments because there's only one screw holding each blade to the hub. Get them fairly close (within a quarter inch or so) and then put another screw into each blade to hold them in position.

Figure 16.6—Aligning and setting the blades with a tape measure.

Carefully turn the rotor over as shown in Figure 16.7. Mount the back plywood hub (Figure 16.8). Use the drill bit through the center and also a square to be sure the back hub is both on center and perfectly aligned with the front hub.

Figure 16.7—Blades flipped over.

Figure 16.8—Setting the back hub with the alignment pin.

Rotate the hub so the screw holes are centered over the blades, as shown in Figure 16.9. Put a screw through the back hub into each blade (Figure 16.10). At this point, if you never care to disassemble the rotor again before installing it on the alternator you could put all the screws in. We prefer to get everything drilled out and ready, then disassemble everything and put finish onto each part separately. For that reason we only put two screws into each blade. An assembled blade set is also difficult to transport, so if you must drive the blades to your site it might be wise to do the final assembly on site.

Figure 16.9—Rotate the hub until the screw holes are centered on the blades.

Figure 16.10—Attaching the blades.

Figure 16.11—Steel hub used as drilling template.

Use the front steel blade hub that you fabricated in the *Frame* chapter (the 6 inch diameter disk cut from 1/4 inch thick steel, drilled out with four holes on a 4 inch diameter to fit the wheel hub, shown in Figure 16.11) as a template for drilling out the holes in the blade for mounting. You could also just lay the holes out carefully by hand, but it's much easier to have a template that you can center on the blades.

Drill the holes for the alternator studs through the blade assembly with a 9/16 inch drill bit (Figure 16.12).

Figure 16.12—Drilling the holes for the alternator studs.

Figure 16.13—George drilling the reinforcement bolt holes through both hubs and each blade.

Wood screws alone will hold the blades together fine, but we like to drill six 5/16 inch holes through the assembly so that two 5/16 inch bolts can be installed through each blade, squeezing the "sandwich" together. This will become more clear in later images of the assembly. George is demonstrating this in Figure 16.13 at left.

Turn the blades over and use a 2-1/2 inch diameter hole saw to cut a hole through the back hub, and all the way through the blades (Figure 16.14). Don't cut into the front hub.

Figure 16.14—Cutting the center hole into the back hub and blades. Note the well-behaved, napping Borzoi shop dog in the background!

You don't need to disassemble everything again, but we prefer to since it makes applying finish easier, and you can be assured that all wooden parts are well-coated with finish. In Figure 16.15 we've removed the front hub. If you do this, be sure to number or mark everything so that it goes back together the same way in the future. This whole assembly only fits together one way. We usually mark each blade with a number and then number the hub.

Now is the time to put finish on the blades. We usually put on a thick coat of linseed oil (Figure 16.16). Linseed oil makes a nice, durable finish, but it requires maintenance. Before raising the wind tur-

Figure 16.15—Rotor parts labelled so you can get them back together correctly after applying wood finish.

bine we apply about four thick coats of linseed oil. After this, it should be wiped down once or (preferably) twice a year. The hubs should also be finished on both sides.

Figure 16.16—Applying linseed oil finish to the blades.

We usually stain them dark (for appearance only) and then coat them with linseed oil.

Insert the six carriage bolts through the front of the front hub (the front hub faces the wind, and is the one without the big hole in the middle), as shown in Figure 16.17.

Figure 16.17—Carriage bolts installed in front hub.

Turn the hub upside down (so the threads are poking up at you) and put the wind turbine blades on facing down so that the bolts go through their holes in the blades (Figure 16.18). The flat surface of each blade that faces the wind should be facing down.

Figure 16.18—Hub upside down with bolts sticking up.

Figure 16.19—Blades back in place.

Figure 16.19 shows all the blades placed over the front hub with the bolt threads sticking up, ready for the back hub to be put on. Notice the letters on each blade to assure that you can get the back hub aligned properly.

Figure 16.20—Tighten all the bolts, then install a wood screw in each of the 60 holes.

Put a washer and a lock washer on each bolt, and tighten (Figure 16.20). You want to tighten these fairly well, so that you'll pull the carriage bolt all the way down into the front plywood hub. Over time the wood will crush a bit, so you'll want to check and tighten these bolts at least once a year. Once the bolts are tight, put a wood screw into each and every hole on the front and back hubs. After you've done that, the rotor is fully assembled and ready to mount on the wind turbine!

Figure 16.21—Rotor attached to alternator, resting on the assembly stand. Do not *ever* let the machine spin in the wind while on the stand! It's extremely dangerous and could get away from you and overspeed before you realize what's happening.

If your alternator is now sitting on your assembly stand in the yard, it might seem really tempting to mount the rotor on it (Figure 16.21). That's just fine, and then you can make sure everything fits properly. However, *do not* allow the machine to spin in the wind on the assembly stand! It will be free wheeling since the alternator does not have a load on it, and could spin up to lethal velocity before you realize the danger. We've seen friends injured and blades wrecked from doing just that—*the only place your turbine should be allowed to spin is on the tower!* If you want to display it in your yard before raising so your neighbors can see what you've been working on in your garage all winter, short all three alternator terminals out with jumper cables, and tie one of the blades to the assembly stand with a rope.

17. Towers

Figure 17.1—A big tower for a big homebrew turbine!

The many hours of research, planning and construction you've been spending to build your wind turbine have probably been a lot of very challenging fun. But the turbine won't do you any good at all sitting on a test stand in your back yard or garage. You need to fly it high on a tower, up where the smooth, concentrated "fuel" of moving air is. As we discussed earlier in this book (Chapter 1, under *Do you have a good site?*), flying a wind turbine on a low tower surrounded by obstructions is like mounting solar panels in the shade.

We considered putting this *Towers* chapter first in the construction sections of this book! The "tower problem" happens frequently—beautiful homebrew wind turbines sit on their test stand in the corner of someone's garage for years without making a single watt, because the builder underestimated the cost and labor involved for erecting a sturdy tower. It's not a bad idea to build your tower first, then start wind turbine construction—that way you can proceed directly to raising and flying the thing when your long labor is done. Or, for every major step in turbine construction you complete, take a break and complete a major step in preparing your site and fabricating your tower.

A good tower takes a lot of money and work. You'll be digging holes in the dirt, pouring concrete, drilling holes into solid rock, or all of the above. As we mentioned in Chapter 1, after you factor in all the costs your tower should cost at least as much as your wind turbine, and frequently twice as much or more. A flimsy tower with thin guy wire anchors is likely to fail during high winds or while being raised and lowered. Besides destroying your turbine, it could mash a building or fall on someone and kill them. Think of you as the fly, and the wind turbine and tower as the fly swatter! Do not underestimate the forces on the tower, guy wires and guy anchors that support even this "small" 10-foot wind turbine.

In this chapter, we'll cover the basic tower types and their advantages and disadvantages. The exact location of your tower site will play a big part in what type of tower you select, and how tall it has to be to get your turbine 30 feet above any obstacle within 300 feet. We'll discuss some commercial tower designs you can either purchase directly or use in your research, and show a good, simple homebrew tower design that you could possibly adapt for your own site. It's highly unlikely that you'll be able to use any of our tower suggestions exactly as printed, since individual sites vary so much. It's up to *you* to do your homework and build a tower that's solid and safe.

Purchase a tower, or design and build your own?

A good solution for installing your first wind turbine is to buy your first tower kit instead of designing and building it yourself. Such a kit will come with all the needed parts—such as couplers between the tower sections, guy wires, guy wire equalizers, thimbles, cable clamps, tower base, hinge, gin pole and approved engineer's drawings to make your friendly local building inspector happy. If it's a tilt-up tube tower, you'll most likely have to supply the steel tubing or pipe yourself. We list many sources of tower kits in the *Sources* section of this book.

Even if you buy a tower kit, you'll still have to provide the tower stub and adapt it to fit the tower top. And you'll have to provide a sturdy foundation and sturdy guy wire anchors. If you do tackle the problem of designing your own tower, there are some resources available. Most tower kit manufacturers let you download the entire specification sheets for their towers for free on the internet. Get all of these that you can, and read them thoroughly. A tower designed for a commercial 10-foot wind turbine will be very close to what you need for a homebrew 10-foot wind turbine—the forces on the tower will be very similar.

If you decide to tackle the project of building your own tower, tower kits are still useful. Download the literature and installation guides for a few of them, and read them thoroughly. This will give you many ideas about how everything fits together, and how sturdy it all needs to be. Some wind turbine seminars also include lessons on tower work, and hands-on experience is the very best way to learn about towers.

If you have a local renewable energy installer that does wind turbines, ask if you can join the crew for a day for free to help on a tower and turbine install. They may be grateful to have extra eyes and hands during the somewhat dangerous process of erecting a tower, and you'll gain invaluable knowledge for your own tower design and installation.

Site and height

We discussed sites and heights in Chapter 1, so just keep in mind that wind speeds are much slower near the ground, and any obstructions nearby form a "pool" of slow-moving air in all directions from the obstacle—not just in front of it. In addition, both the ground and obstacles cause turbulence, which lowers power output and puts unnecessary stress on the machine. A good rule of thumb is that any wind turbine should be flown at least 30 feet above any obstacle (in any direction) within 300 feet. Many wind power experts recommend even taller towers and more distance from obstructions than that. Don't skimp on tower height—it's the most important thing you can do to assure good power output and long life of the wind turbine.

The terrain around your tower site will in some ways dictate the best tower design for you. We discuss the advantages of each tower type on different sorts of terrain below, under "tower types."

GOVERNMENT ORDINANCES

Many local governments have restrictions on tower heights. Above a certain height, you may need a special permit, which will often entail submitting certified engineer's drawings of the tower. Most local regulations require that if the tower should fall, all of it would land entirely on your property. If you live close to an airport, you may have to check with the Federal Aircraft Administration (FAA) for tower restrictions and lighting requirements—however, if you are more than 4 miles from an airport no special permission is required from the FAA unless the tower is 200 feet tall or over. In general, the more rural and remote your area is, the fewer regulatory hassles you'll have to endure in getting permission to erect your tower.

You should check into all of this before even considering building or buying a wind turbine. Many cities won't allow tall towers at all. This has led to the promotion of "rooftop wind turbines,'" which are a very bad idea (see

Chapter 1)—rooftop towers are not generally high enough to get above obstructions, they put extreme forces on your house that it was not designed to take, and the vibration from the wind turbine will likely be very irritating during a wind storm. Even quiet wind turbines like this one transmit vibration to the tower when they are generating power, and you'll be able to hear and feel this vibration quite clearly. We strongly urge you to not even consider a rooftop wind turbine tower. If you do build a rooftop mount, don't come crying to us if your windmill makes very little power, keeps you awake all night, or causes your house to fall down!

NEIGHBOR RELATIONS

You'll want to discuss your wind turbine and tower project in advance with any neighbors who are near enough that they'll be able to see your turbine. Again, we discussed this in Chapter 1, but it's worth mentioning again. Your neighbors may be concerned about noise and visual aesthetics. It might pay off in the neighbor relations department to paint your tower and wind turbine white or light blue to lessen their visibility against the sky. Strobing can be an issue, where for a couple weeks twice a year, the sun will be directly behind the spinning blades for a few minutes, casting a moving, flickering shadow on their home. Noise is generally not an issue with this wind turbine design, it's one of the quietest turbines we've heard. We generally describe the noise as being similar to someone riding by outside on a bicycle, and usually the wind in the nearby trees is much louder than the turbine itself. Tower vibrations can also cause noise, but this can usually be remedied by small modifications to the tower structure to prevent rattles.

Tower types

There are many different tower types that can be used for wind turbines. However, we class them into two main groups—the kind you have to climb, and the kind that tilts down to the ground. If you choose a tower that must be climbed, you must have the proper training, physical conditioning, and safety equipment to feel confident hanging high in the air like a piñata while installing and maintaining your wind turbine. Keep in mind that any wind turbine will need regular maintenance at least once a year!

The geography of the site you are planning on may also dictate the tower type needed. Each tower type described below also includes a drawing of the tower's "footprint" on the ground—the amount of free space (cleared of all structures, trees and brush) that is needed, and the relative size of the concrete that needs to be poured.

FREESTANDING

Freestanding towers have the big advantages of needing no guy wires and of having the smallest footprint of any tower type. This also makes them very suitable for rough or heavily forested sites. However, these towers do need an extremely large chunk of concrete at the base to make up for the lack of guy wires. They also need to be very robust, built with lots of steel, and are generally very expensive. The radius of the concrete base for a freestanding tower will have to be at least 7–10 percent of the tower height.

Figure 17.2—A utility-scale wind turbine on a monopole tower.

The footprint of a typical freestanding tower is shown in Figure 17.4. The two primary types of freestanding towers are monopole (Figure 17.2) and lattice (Figure 17.3). Neither type of freestanding tower is suitable for designing and building at home, they must be purchased commercially. And, a crane service is usually required for installing or removing the tower and turbine. Freestanding towers generally must be climbed, there are few tilt-down versions.

Figure 17.4—Footprint of a typical freestanding tower. The dotted line shows the area that must be cleared of trees and brush, the hatched area is concrete, and the black area is the tower itself.

Figure 17.3—A freestanding lattice tower.

Some people have used old, freestanding farm windmill towers for their new electricity-producing turbines. This is possible—but remember that your life depends on a very careful inspection of the integrity of the old tower.

Any deterioration caused by corrosion is an extremely dangerous situation, as the tower could buckle under you while you are climbing it. And, such corrosion might be below grade, where you can't see it. If in doubt, contact an expert before climbing an old tower!

Check the *Sources* section of this manual for information on where freestanding towers can be purchased. The manufacturer will provide information about how large the tower must be compared to the wind turbine swept area, the size of the crane and crew needed, the specifications for the concrete base versus the tower height, and so on.

GUYED LATTICE

Figure 17.5—A guyed lattice tower.

These towers are quite common, and are often used for wind turbine installations (Figure 17.5 at left). While lattice towers are once again not something you can successfully build at home, they are often available used from ham radio operators. Lattice towers also must be climbed, and they require three sets of guy wires and three guy wire anchors. Generally, a set of guy wires is attached where each tower section joins the rest (usually every 10 or 20 feet). The guy radius of a lattice tower will usually be from 50 to 80 percent of the tower height—check with the manufacturer for the exact specifications needed.

The footprint of a guyed lattice tower is shown in Figure 17.6, facing page. Hiring a crane service makes erecting a guyed lattice tower into a one-day proposition—it's often done in only a couple "lifts" with the crane, even if there are a dozen tower sections to go up.

But it is possible for a small crew to erect one on a remote site without a crane, using a daring climber, some exotic hardware, and whole lot of concentration and communication.

To do this, a pole with a davit (a small, homemade crane that bolts to the tower) and pulley on top is clamped to the tower frame, and the next tower section is hauled up by a

---

CRUCIAL DEFINITIONS

Guy radius
The distance from the center of a tower to the guy wire anchors. Usually expressed as a percentage of the tower height.

rope from the ground. After the section is bolted in, the climber moves the davit up to the new section, and hauls up the next tower section and bolts it in. And so on, until the final load hauled up on the davit is the wind turbine itself. This is not something for the inexperienced tower climber to attempt.

Be sure to do thorough research and contact the tower company before choosing your tower—some suppliers of lattice towers are covered in the *Sources* section. Lattice towers are available in a variety of sizes (measured by the distance between the three legs and the size of the pipe on each leg), and too narrow a tower can buckle. Do your research in advance, or hire a qualified wind turbine and tower dealer/installer!

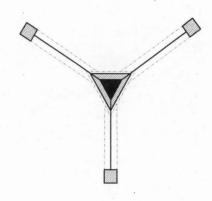

Figure 17.6—The footprint of a guyed lattice tower.

Tilt-up tube towers

The authors don't climb any wind turbine towers on a regular basis. We always use tilt-ups to fly our own turbines, no matter how difficult the terrain is—we'd rather deal with the engineering challenges involved than hang high up in the air. When dangling from a tower top, you also need to be extremely mindful of watching the skies for giant pterodactyls, which can pluck you off the tower and eat you with no warning. You could also ask yourself this question: "Would I risk my life flying in an airplane that I designed and built?" Our answer would certainly be "heck no"—the duct tape holding the wings on would degrade in the sun, the baling wire

Figure 17.7—A tilt-up tower halfway raised.

securing the propeller might unravel, and the chewing gum holding in the engine could stretch to the breaking point! So for the remainder of this chapter on towers, we'll be focusing entirely on the tilt-up variety.

Tilt-up towers give you the huge advantage that all your work on the turbine (including installation and both scheduled and unscheduled mainten-

Figure 17.8—Maintaining a wind turbine while standing safely on the ground. We love tilt-up towers for this reason!

ance) are performed while you are standing safely on the ground (Figure 17.8). Once the tower is installed and adjusted, raising and lowering is a simple matter of shutting down the turbine and using a winch or vehicle to slowly raise or lower the tower and turbine. Don't underestimate the danger of tilt-up towers, either—some experts consider erecting them more dangerous than climbing fixed towers!

Instead of the three guy wire sets used in lattice towers, tilt-up towers use four sets. This configuration keeps the tower stable while it tilts up and down. A tilt-up tube tower is shown halfway up in Figure 17.7 (previous page). The guy radius of a tilt-up tower is usually between 35 and 60 percent of the tower height. We use a guy radius of 50 percent, and that way there's no extra guy wire needed from the tower top to the guy anchor. Instead, the end of the gin pole on the "raising" side is attached directly to the guy anchor once the tower is erect, and the wire rope between the gin pole and tower top serves as the guy wire for that side of the tower.

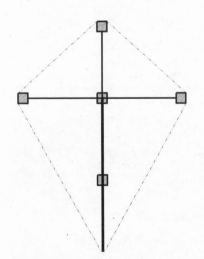

Figure 17.9—Footprint of a tilt-up tower. It needs the largest cleared footprint of any tower type.

The big disadvantage of tilt-up towers is the large footprint on the ground that they take up (Figure 17.9). The entire area within the dotted lines must be completely cleared of any trees or brush. And, the more uneven the ground, the more difficult the tower will be to raise and lower. The ideal site for a tilt-up tower would be a perfectly flat and level field—which is unfortunately not something we have in abundance here in the Rocky Mountains. Fortunately, there are ways to compensate for difficult terrain with a tilt-up tower, and we discuss them a bit later in this chapter.

GIN POLE

The key to raising and lowering a tilt-up tower is the "gin pole"—it provides the leverage needed in the correct direction, and is shown clearly in Figure 17.7 on the left side of the tower. The gin pole will often have a pulley (Figure 17.10, next page) installed at the top, to provide a mechanical advantage. But remember that this 2:1 advantage also means you'll have to spool in and out twice the amount of cable to raise and lower the tower.

RAISING AND LOWERING

There are a few different ways to get your tower into and out of the air. All of them require a mechanical advantage—the turbine and tower are far too heavy to just push up into the air, even with a whole group of people.

Winch: A permanently installed winch (Figure 17.11) is a great way to slowly and gently move a tower up and down. It can be driven by its own electric motor, with a cordless drill, or with a hand crank.

Vehicle: A very slow-moving and powerful vehicle can be used to raise a tower. A tractor or a 4-wheel-drive truck with a low range transfer case are preferred—this is not a "floor it and pop the clutch" operation! Any bouncing or jerking can overstress the tower and cause a disaster. You'll need a long stretch of smooth open ground for the vehicle, and even longer if you use a pulley at the top for a slower ascent. The driver should face the tower, *not* away from it, for safety.

Figure 17.10—A pulley at the end of a gin pole, to give a mechanical advantage while raising and lowering.

Figure 17.11—A big electric winch for raising and lowering the tower.

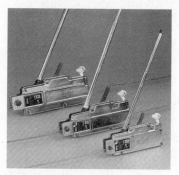

Figure 17.12—Griphoist® wire rope pullers.
Photo courtesy of Tractel Group.

Griphoist®: These handy tools (shown in Figure 17.12) are like a ratcheting "come-along," but since they "walk" the cable, there is no length limit on the amount of cable they can pull through. They are very effective if your tower site has no electricity or vehicle access.

Block and tackle: If you can gather 5–6 people to help, rope and a block and tackle can be used to raise a tower (Figure 17.13).

Figure 17.13—There are no vehicles or winches in the town of Set Net Point, Nicaragua...but a block and tackle and six willing volunteers does the trick.

TUBING OR PIPE?

Steel pipe or tubing can both work well for a tilt up tower. Pipe and tubing look very much alike, but there are differences you should know about if you purchase one or the other. Pipe is measured by inner diameter and the wall thickness is rated by "schedule." Schedule 40 pipe is the most common, and the higher the number the thicker the walls. Pipe is normally provided in 21-foot lengths. Tubing is measured by outer diameter, and the wall thickness is measured in gauge just like sheet metal. With tubing, the lower the number, the thicker the walls. Tubing is normally provided in 20-foot lengths. The strength of a tower built from tubing or pipe is related directly to the wall thickness, and to the square of the diameter—double the diameter and the tower is 4 times stronger. Seamless pipe of any size will fit tightly over tubing of any gauge, in the same size.

The minimum size of pipe to safely support a 10-foot wind turbine is 2-1/2 inch schedule 40 steel. We've started using 12 gauge tubing in 6 inch diameter instead of pipe—a larger diameter gives a bigger increase in strength for less cost than a thicker pipe wall. It doesn't hurt to go larger, as it's better to overbuild your tower than underbuild it. Oftentimes 25 percent more cost can mean twice the strength. One problem with using tubing instead of pipe is availability—if you are not near a good steel dealer, you may have difficulty locating large diameter steel tubing, in which case pipe will work fine. For a 40-foot tower, we have successfully used a 21-foot section of 3 inch pipe, with another section of 2-1/2 inch pipe above, nested inside the 3 inch and overlapping by 2 feet. Or, you can use all 2-1/2 pipe, with connectors either inside the pipe, or wrapped around it.

Figure 17.14—Guy wire mount for the top tower section.

Never use threaded connectors anywhere on the tower, it's a major weakness! A threaded pipe (schedule 40) has less than 50 percent of its wall thickness at the bottom of the threads, and less than 1/4 of its original strength. It's much better to nest the pieces of pipe inside each other, or to build couplers that clamp the pipe together from the inside, around the outside, or both. In fact, that's the big advantage of buying a commercial tower kit—the couplers for the pipe section are included and well-designed. The couplers are the most complicated part of the tower to fabricate. Figure 17.14 shows a top guy wire mount, and Figure 17.21 shows a coupler. Each coupler is attached to a set of guy wires.

LAYING OUT THE TOWER FOUNDATION

When you hire a carpenter to build your house, you expect the final framed result to be square, level and plumb. There are good reasons for this—any part of the structure that's not both level and plumb will have a drastically reduced capability for supporting a large amount of weight. And anything that's not square will cause you problems over and over again as you finish the inside of your house—drywall will have to be cut at crazy angles, cabinets won't fit, and the folks installing your granite countertops will angrily ask you "How the #$^#&! did you come up with a 94.7 degree angle here?" while adding a few hundred dollars extra to your bill. Wind turbine towers are *exactly* the same!

Your tower will be supporting a tremendous amount of weight, and if the tower is not plumb, you risk it crumpling to the ground. If your guy wire anchors are not perfectly square with each other, the tower will be a huge hassle every time you need to raise and lower it—the side guy wires will tighten or loosen too much for the turnbuckles to compensate for as you raise and lower the tower, making for a tedious and dangerous operation every single time the turbine needs regular maintenance. If your tower base and guy wire anchors are so out of alignment that the turnbuckles can't take in or let out enough slack, you might as well just start over and lay out your tower again, from the beginning.

There's a way to align the side guy wire anchors slightly offset that can really help your tower raising operations. This offset assures that the guy wires loosen when the tower is down, and tighten as it goes up. It doesn't take much—we advise trying one inch of offset on the side guy wires toward the tilt-down end of the tower, opposite the axis of the tower hinge, for every 10 feet of tower height. Keep in mind how little tolerance this is. Your guy wire anchors and base should be aligned very precisely before you pour in tons of concrete—they will last for hundreds of years, whether you get them in the right spot or not!

And, if you must install your tilt-up tower on slanted ground, be sure the turbine tilts down towards the uphill side of the site. Otherwise you'll have to modify the yaw and tail bearings so that the whole machine doesn't just slide off the tower stub and crash to the ground when lowered.

Figure 17.15—A simple tower base and hinge.

TOWER BASE AND HINGE

With a tilt-up tower, the base takes the least amount of abuse. All the force on the base is simply the weight of the tower and turbine, pushing straight down. Some commercial tilt-up tower kits actually don't even specify concrete for the tower base, they simply have a very wide, spread-out metal base that is spiked into the ground with long pieces of rebar. We prefer to pour a concrete base and fasten the tower base and hinge to it with bolts cast into the concrete, which gives a more stable hinge. A simple tower base and hinge for a 10-foot machine on a 40-foot, 2-1/2 inch schedule 40 pipe tower is shown in Figure 17.15.

GUY WIRE ANCHORS

After the wind turbine and tilt-up tower are erected, the guy wires and their anchors take most of the stress from the wind. The standard method of installing guy wire anchors used by utilities and most wind turbine tower kits is to auger a 6 inch diameter hole in the dirt and install "expanding earth anchors" or screw anchors (see the *Sources* section). However, you must ensure that your soil type is correct or these anchors will not hold. They are also rather expensive, and you'll need power equipment to auger the hole deep enough.

Figure 17.16—Guy wire anchor embedded in concrete.

Another option is to pour a whole bunch of concrete, and embed steel bars with barbs so that the ends stick out a bit from the concrete. (Figure 17.16). Each of the four guy anchors in a tilt-up tower connects with all the guy wires from each side of the tower. The gin pole is also guyed to the side guy anchors, to keep the tower from trying to flop over to the side (and breaking the base hinge) during raising and lowering.

A large "dead man" can also work in difficult terrain—this is usually a long piece of pipe or other big chunk of steel, covered with lots of dirt and/or concrete. One of co-author Dan B's guy wire anchors is a large old, non-functional gasoline generator dumped in a hole with concrete and dirt covering it. Besides making a dandy guy wire anchor, he got rid of a piece of junk cluttering up the yard!

Sometimes it's not possible to dig large holes and pour sturdy concrete guy anchors with embedded steel plates, or to auger deep holes in the dirt—there's a reason they call them the "Rocky Mountains" up here where we live. In some cases, if the rock formation is large and strong enough, it's possible to drill out solid rocks (Figure 17.17) and pound in large expansion bolts for guy anchors. A homebrew option is to pound #6 or #8 rebar

Figure 17.17—Rich drilling out a rock formation for an expanding bolt guy wire anchor.

with loops welded to one end into the holes, after first injecting epoxy into the hole—but expansion bolts are much preferred.

WIRE ROPE AND HEAVY HARDWARE

All of your guy wires will be made of wire rope, with a minimum guy wire size (for a 40-foot tower and this 10-foot turbine) of 1/4 inch wire rope. Each guy wire will need thimbles anywhere it connects to the tower or the guy anchors, large turnbuckles (no hooks allowed, solid loops on the ends of the turnbuckles only), and exactly 3 properly-oriented wire rope clips per connection. And remember, "never saddle a dead horse!" Don't know what any of this talk about thimbles, turnbuckles and dead horses means? Then you need to do

RIGGING TERMINOLOGY

Expanding anchor bolt: Used to secure guy wires into rocks or concrete.

Shackle: Used to connect wire rope to guy anchors, towers, etc.

Thimble: Protects the interior of a wire rope loop from abrasion.

Turnbuckle: Allows you to adjust guy wire length. The top turnbuckle pictured is a "jaw to jaw" model, the lower one is an "eye to eye." Never use turnbuckles with hooks—jaws or eyes only!

Wire rope clip: Fastens one length of wire rope to another. Used to makes loops for attaching guy wires. A thimble is inserted into the wire rope loop, which is held by 3 clips.

NOTABLE QUOTES

"Never saddle a dead horse."
Author unknown

This is the rule for using wire rope clips. The saddle of the clip should always grasp the live end of the wire rope, not the dead end.

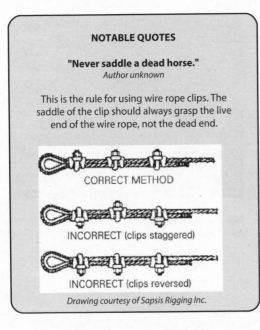

CORRECT METHOD

INCORRECT (clips staggered)

INCORRECT (clips reversed)

Drawing courtesy of Sapsis Rigging Inc.

your homework on towers and wire rope rigging! We give you a head start in the sidebars on this page and the previous page.

Homebrew tower designs

All right—you are still reading, so obviously we haven't convinced you to purchase your first tower as a kit! That's OK, but we'd still like to emphasize: Find a local wind turbine installer or neighbor who will demonstrate for you how a tilt-up tower works. Or, sign up for a wind turbine seminar where tower fabrication and raising is included. There are some tricky aspects of tilting a tower up and down that are confusing to read about, but will become perfectly clear when you see everything in action. And, a fun tower project idea you can tackle on the kitchen table is to build an exact scale model of your tower idea out of wooden dowels and string. Then you can safely visualize exactly what will be happening when you build the real one out of steel.

There are only a couple of homebrew tilt-up tower designs that we are comfortable with. And the *only* thing we can assure you of about the single tower design we present here is that the tower itself is strong enough to support

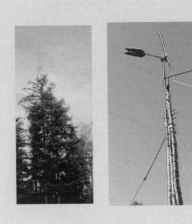

100 percent organic towers?

Well, mostly organic. And yes, we and others have made them. Using trees, either living or cut, might seem like a cheap solution for building a wind turbine tower, but we don't recommend it! Too many things can go wrong. The tower on the left is pinned to a living, 120-foot tall spruce tree. The owner and installer is a professional arborist and climber, and knows what he's doing—and knows how to keep the tree alive. As for the tower on the right...let's just say that we really needed some electricity fast, and strapping these trees together with tire chains was our only option!

the wind turbine, and that the gin pole is long and strong enough to raise it! Everything else—base, hinge, guy wires, guy wire anchors—is your responsibility. It's always better to overbuild than to underbuild.

OUR FAVORITE TOWER

This one is becoming a standard up here (Figure 17.18). We've moved almost completely to building towers from 12 gauge steel tubing instead of schedule 40 steel pipe. The tubing is stronger by weight, and easier to work with. We even use larger versions of this tower (10 inch tubing instead of 6 inch) for our big 17-foot wind turbines (Chapter 20, *Scaling it up and down*). The main bits of fabrication you'll have to do are the section couplers, the tower top stub, and the hinged base.

Figure 17.18—Our favorite wind turbine tower for the 10 footer, made of 6 inch diameter 12 gauge steel tubing.

The tower is 60 feet tall, not counting the 4 foot stub on the top where the turbine is mounted. It's made from 6 inch diameter 12 gauge steel tubing. You could easily make this tower taller by adding additional tubing sections, couplers and guy wires. 100 feet would be about the safe maximum height. The tower is built from 3 sections (60 feet) of tubing. The gin pole is 30 feet long, built from the same tubing. The 6 feet of tubing that remain are used for couplers.

The base of the tower is fabricated from angle iron and pipe (Figure 17.19). In the parts list (below) we don't list exact sizes of these steel bits, because we usually just pick and choose from what's available at the site or at our shop. The gin pole and the tower will slip into the 6 inch diameter pipe, which is welded to a pivot so the tower can tip up and down. The whole tower will pivot on a 3 inch piece of pipe (the axis) and the bearing is 4 inch diameter pipe. You don't want a "tight" fit in the bearing. Since the base could never hold the tower up without guy wires, we like to have a good bit of play in the pivot. 4 inch pipe over 3 pipe works nicely in this application. The whole base is about 3 feet square.

Figure 17.19—Tower base for the 6 inch tube tower.

Figure 17.20—The tower top stub that must be fabricated for mounting the wind turbine.

The tower stub (the part that the wind turbine fits onto) is also built from pipe, and is shown in Figure 17.20. The bottom sleeve is 8 inches long, made from 6 inch diameter schedule 40 seamless pipe, and it fits tightly over the tubing from which the tower is fabricated. On top of the 6 inch pipe we weld a 8 inch diameter steel disc (cut from 1/4 inch steel). On top of that we weld the stub, which is 18 inches long and made from 2 inch schedule 40 pipe. The yaw bearing for the wind turbine is 2-1/2 inch pipe, and the fit between it and the 2 inch pipe is good—not too tight, not too loose. On top of the tower stub we weld a ring 2 inches outer diameter (this just fits inside the 2-1/2 inch pipe) and 1 inch inner diameter. This gives some surface area at the top and acts as somewhat of a bushing. Without this, in a short time the tower stub would wear into the top of the yaw bearing. That could make it hard for the machine to yaw and—even worse—furl.

Figure 17.21—Closeup of a gin pole coupler, with no guy wire attachments.

Figure 17.21 shows a gin pole coupler, while Figure 17.14 shows a tower-top guy wire mount. They are all similar, except the couplers for the tower have attachments for guy wires and the ones for the gin pole do not. There are two for the tower, one for the gin pole. They're each 2 feet long. We split the 6 inch tubing and weld bar stock to it with holes drilled so that they can be tightly clamped around the tubing with 3/8 inch bolts. The guy wire connections are just 2 triangles cut from 1/4 inch steel with 3/8 inch holes so that the guy wire loops around a 3/8 inch bolt.

The overall length of the gin pole is 30 feet, which is also the same as the guy wire radius. The end of the gin pole has accommodation for two guy wires to be anchored, and a large pulley at the end. The cable we use to pull the tower up with will be anchored right beneath the end of the gin pole, go through the pulley there, back down to the ground (also very near the end of the gin pole) through another pulley—and then finally to a winch permanently mounted in the concrete.

Once the gin pole is assembled, it should be guyed just like the tower, so that it cannot fall over. The tower base and hinge are not strong enough to support the gin pole without guy wires. The guy wires that go from the gin pole to the tower should also be attached with turnbuckles (once the gin pole is raised it's too late). Oftentimes a temporary gin pole should be built to raise the real gin pole (yes, a gin pole for your gin pole!), or a backhoe or tractor with a bucket can be used to pull the gin pole up. Once the gin pole is raised, you can start assembling the tower pieces right onto the base hinge sleeve.

Carry each tower tubing section into position in the coupler, and prop up the far end to keep it aligned while tightening the coupler bolts. After each section is bolted securely in place, move to the next one (Figure 17.22). The tower top stub is the last section to be secured in place.

60-FOOT TILT-UP TUBE TOWER PARTS LIST

- 6 inch diameter 12 gauge steel tubing, 20 foot sections, quantity 5

- 6 inch diameter schedule 40 steel pipe, 2 feet

- Miscellaneous 1/4 inch thick steel angle and channel bits, 2-1/2 to 4 inch width, for base hinge frame and tubing coupler reinforcement

- Miscellaneous 2, 2-1/2 and 3 inch schedule 40 pipe bits, for base hinge and tower stub

- 1/4 inch wire rope, 500 feet total

- Turnbuckles, 1/2 inch / 12 inch jaw, industrial grade forged eye, quantity 8

Figure 17.22—Assembling the tower pieces. In this photo, Rich is pulling the power output wires through each section as they are assembled.

- Thimbles, 1/4 inch, quantity 16

- Cable clamps, 1/4 inch, quantity 60 (each connection has 3 clamps)

- Shackles, 1/2 inch, quantity 10

- Guy wire anchors. Could be any variety of materials

Test raise

With a tilt-up tower it's an extremely good idea to raise and lower your tower a couple times before you install the wind turbine on top of it. Once you get the whole process going smoothly, your personal stress level during the actual erection of the turbine will be minimized. Before beginning the test raise, check and then re-check *all* of your cable clamps, turnbuckles and guy wires. Lay the guy wires out neatly on the ground so nothing will tangle as things go up. Doing multiple test raises before attaching the turbine is always a good idea, and after you get the guy wire lengths right, each raise won't take much time. We often do a test raise, then add the next tower section, then raise again, and so on until the tower top stub goes up on the last test raise.

In Figure 17.23, Rich (at left) is pulling on a rope attached alongside the top guy wires. When raising a tower, someone needs to pull on this rope as the tower approaches nearly vertical, to overcome the weight of the gin pole—otherwise at a certain point the gin pole side of the tower will drop rapidly to the ground, making for a rather violent erection.

Figure 17.23—Test raise of a tower. At left, Rich is ready to pull on the rope used to overcome the weight of the gin pole.

The test raise is always a bit scary. We try to figure our guy wire lengths as accurately as possible, but usually we cut and adjust them so the tower is slack on the first raise, and then tighten them a bit. Remember that tower guy wires should *always* tighten as it goes up and go slack a bit when it goes down. The turnbuckles will only take up or release a limited amount of slack during this process—if you need to take up more than the turnbuckles allow, you'll have to sequentially un-fasten and re-fasten all the cable clamps on each guy wire to re-adjust the length. As you can imagine, this is much less stressful when there's nothing valuable perched on the tower top. After a few test runs, you'll find a happy medium for where your guy wire lengths and turnbuckles are set so that they need no adjustment at all during raising and lowering.

The problem is that you can't tell how much stress a wire rope is under by looking at it. It could have 30 pounds of force on it or 3,000 pounds, and you'll never know the difference. Pay close attention to guy wire tension during your test raises, by physically touching them. They should not be hanging slack (though they might during the very first raise), nor should they be tight as guitar strings. This is yet another reason why we recommend that you consult with an experienced wind turbine installer before you try your first tower.

The bottom line about towers

Towers for wind turbines are big, heavy, high and dangerous. Either know what you're doing or find somebody who does, if it comes down to engineering, fabricating and installing your own tower! Otherwise buy a properly engineered tower kit.

Novel ways to increase your average wind speed

We discussed tower height and average wind speed in Chapter 1, *Introduction to wind power*, and included a dandy graph of the effects of tower height in Appendix F, *Useful wind data*. But there are some other innovative techniques we've developed over years of trial and error (mostly error) to increase average wind speed at any given wind turbine site.

• Of course, a taller tower is the best way to increase the average wind speed that your turbine sees.

• Try loosening some of the bolts on your solar panel mounting frames. This almost guarantees a large increase in wind speed, usually enough to rip a few of your expensive solar panels off the roof and fling them to the ground, leaving shards of broken glass everywhere.

• Shade umbrellas for the patio furniture on your deck can also be effective for increasing wind speed. The more umbrellas you erect and the more expensive your furniture is, the greater the wind speed increase you'll see as your tables and benches become airborne and float serenely over the valley far below.

• If you are currently building your home, constructing the roof often increases average wind speed at your site dramatically. As you wrestle 4x8 sheets of plywood high in the air, the wind will almost certainly increase enough to send you flying as you grasp the sheet, with the plywood acting much like a parasail or hang glider. The higher your roof line, the greater the increase in wind, and the faster the speed rises. Also note that any curve in the plywood sheet you are desperately clutching can create lift (Chapter 3, *Power in the wind*) and give you an awe-inspiring ride. Bring a camera! However, we've never quite figured out the landing part of this technique—it can be very challenging.

• For a sure-fire way to *decrease* average wind speed, though, simply erect a wind turbine! This guarantees that your site will see no wind for a period of 3 days to 3 weeks. We are considering patenting this idea to protect trailer parks from tornadoes. If only one trailer park resident were to erect a small wind turbine each week, most areas would be completely protected for the entire storm season. The larger and more expensive the wind turbine, the greater the decrease in average wind speeds over a longer period of time.

DOG HAIKU

Make your turbine last
Never use a tall tower
Leave it in the box

18. Raising

Well, the big day has arrived! Time to fly that wind turbine you've spent so long building. Actually, there are quite a few things you need to take care of first. Those include:

• Installing your rectifier, breaker, controller and stop switch, and wiring them to your system.

• Running the wires from the turbine to the system.

• Installing the turbine on the tower top.

• Mounting and balancing the rotor.

• Going through the pre-raise checklist thoroughly.

• Raising the wind turbine!

Mounting and wiring the controls

Figure 18.1, reprinted from Chapter 5, *Furling and regulation*, shows the basics of wiring your wind turbine control system to the rest of your power system.

RECTIFIER ASSEMBLY

You'll want to mount the rectifier assembly that you built from Chapter 13, *Rectifiers*, on a vertical wall in a safe area, and *not* in the same enclosure as your battery bank. The rectifier assembly will get warm during high winds, and if things go wrong (lightning, sized wrong, inadequate heat sink, etc.) it could get *hot*, so keep fire safety in mind when you choose a location and the type of surface to which you mount it. Mount it so that the heat sink fins are oriented vertically, as this improves air circulation and cooling.

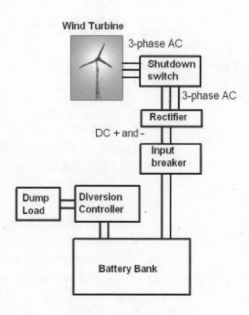

Figure 18.1—System wiring diagram.

When you built the rectifier assembly, did you remember to mount the rectifiers onto the heat sink with a layer of special "heat sink grease" in between? If you slacked on this because it took an extra trip to the electronics store, now is a good time to fix it. The heat sink grease is essential! Without a large heat sink and excellent thermal contact, no rectifier will be able to carry anywhere close to its rated amperage. When you first raise your wind turbine and it's operating, take the time to feel your rectifier's heat sink. On windy days, if the unit is getting "hot" then you likely need a larger heat sink. There's no such thing as a heat sink that's too big—and it should never get much more than "warm to the touch."

STOP SWITCH

You might have installed a stop switch right on your rectifier assembly like we usually do (Chapter 13, *Rectifiers*). If not, now is the time to make your stop switch. With this turbine, all you need to stop the blades quickly is a big switch to short all three incoming phases to each other, somewhere between the wind turbine and the rectifier. You'll use this switch to shut everything down during raising and lowering, for protection during extreme winds, and in case something goes wrong up there on the tower or in your RE system wiring while the wind is blowing. The switch should be a double pole, single throw (DPST) switch rated 30 amps or more, and the schematic of how it should be wired is shown in Figure 18.2.

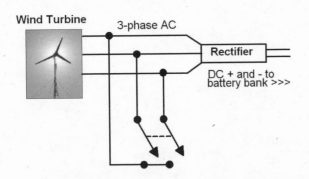

Figure 18.2—Shutdown switch wiring diagram.

INPUT BREAKER

From the rectifiers and stop switch, you should go through an input fuse or breaker. It's also required by the NEC electrical code. This should be quite a lot larger than the maximum current you'll ever see from the wind turbine—if the fuse blows or the breaker trips, the wind turbine will be free spinning with no load on it, which is a dangerous condition. If this ever happens (very unlikely) you should stop the wind turbine with the shutdown switch immediately. We suggest at least a 50 amp fuse or breaker for a 48 volt machine,

100 amps for 24 volt, and 200 amps for a 12 volt machine. Label the breaker clearly, it's not the same as the stop switch! It's more like a "self destruct switch" if accidentally opened by a human during high winds. The most likely reason this breaker would ever trip would be serious problems—a major short circuit in the controller or battery system wiring, which is extremely unlikely.

CONTROLLER

From the rectifier, you are now running DC power to your battery bank on two wires, a large (+) and a large (–) wire. We also discussed controller types in Chapter 5. By now you should have one of the three different kinds of controller ready for installation:

Manual controller: (Running more lights and appliances during high winds, and shutting the turbine down with the stop switch if you can't burn enough power). If this method is what you are going to use, no further installation is required. We don't recommend this, however, as if you forget to shut the turbine down when you leave home, you could come back to a disaster—overcharged batteries and a wind turbine destroyed from overspeeding.

Diversion load controller and heating element dump loads: A properly-sized Xantrex C-series or Morningstar TS series, set for diversion control mode (instead of standard solar panel control mode), and a properly sized array of air or water heating elements (1,500 watts). The recommended controller capacity for each system voltage is: a 130 amp controller for a 12 volt system, a 70 amp controller for a 24 volt system, and a 40 amp controller for a 48 volt system.

Again, notice that for a 12 volt system you may need two controllers wired in parallel, or use the voltage controlled switch method below instead.

As with the rectifier assembly, use care in mounting the diversion controller. It can get quite warm during high winds, and must be mounted on a vertical wall in the orientation specified by the manufacturer for proper cooling. The dump load heaters themselves will get *very* hot (of course!) and should also be mounted with fire prevention in mind, carefully following the manufacturer's instructions.

Aux control relay and heating element dump loads: With this type of controller system, use the same mounting precautions as with the diversion controller above. You'll need to set the hysteresis so that the control relay doesn't cycle on and off too quickly—no faster than once a minute. A good place to start with setting your hysteresis is 1 volt for a 12 volt system, 2 volts for a 24 volt system, and 4 volts for a 48 volt system. Remember that a plain old relay may not cut it for switching that large amount of power—unlike with the diversion controller above, this system will always run the heaters at full blast when they turn on. Use a mercury contactor or solid state relay to switch the load.

Turbine wiring

The wire from the wind turbine to the rectifiers and battery needs to be sized according to the current that the wind turbine might generate, and the distance between the wind turbine and the batteries. The lower the voltage the higher the current—this is yet another reason higher voltage systems are easier to get along with and less expensive. The wire run from the alternator to the tower base is different than the wire from the tower base to the rest of the system. This turbine design doesn't use slip rings to get power from the alternator down the tower; instead we use a simple, easy to build, and more reliable system called a "pendant cable." This wire runs down to the ground inside the tower pipe, and actually twists as the wind turbine yaws to face the wind.

Figure 18.3—Heavy duty locking plug and socket at the tower base.

You should select the best grade available of flexible, stranded #8 or #10 AWG extension cord wire for the tower wiring. At the tower bottom, you'll need to install a heavy duty, locking 3-prong electrical plug (Figure 18.3, at left) on the 3 wires coming down from the turbine, and a matching socket that's connected to the wires. You need to size the plug and socket for the same amperage rating as your controller. When you price the plug and socket at the electrical supply store, you'll once again see how higher-voltage systems save money! We highly recommend a locking plug and socket, as you do not want this connection to come loose accidentally. It's also handy to buy an extra plug and wire it so it's a dead short—insert this into the turbine output plug, and you can stop the machine immediately.

As you'll read later in the section on turbine maintenance, every couple weeks you'll need to check this plug and socket to see if the pendant cable wires have twisted. If they have, you'll simply unplug at the bottom of the tower, untwist the wire, and plug it back in. Insert the extra shorting plug that you made to stop the turbine while you perform this untwisting operation. The more turbulent your tower site is, the more often you'll have to untwist. Tall towers on excellent wind sites might need untwisting only 3–4 times a year, while turbines mounted on too-short towers in bad locations could need this weekly.

From your socket at the tower base to your rectifier, you'll most likely have a long run of wire. This does not have to be flexible, so you can use standard electrical cable of the proper size, directly buried or enclosed in buried conduit.

The lower your system voltage, the thicker this wire will have to be, and the longer your wire run, the thicker your wire will have to be. And the thicker the wire, the more expensive it will be! If you ran the possible maximum output of the wind turbine (1,500 watts) through the wire sizing formulas for solar PV systems (usually for only 5% loss), you'd be told to use extremely thick and expensive wire. Fortunately for your wire budget, the turbine will produce 1,000 watts or over very rarely, since high winds are so rare. And during such windy periods, you'll have full batteries pretty quickly. So you can size the wire for a lower output, and accept more line loss at higher wind speeds.

However, there's a limit to how much loss you can accept. The more power that's wasted, the faster the wind turbine will run and the less efficient the system will be overall. If the wind turbine runs too fast (too much power wasted in the line), the blades will become less efficient. They may also become noisy and the machine may not furl properly. An overspeeding wind turbine is dangerous and one more reason to make sure you size the conductors appropriately. So, the wire size recommendations in the chart below should be considered the minimum thickness of wire to use. It won't harm anything to use thicker wire, but it will be more expensive.

Recommended minimum wire gauge (# AWG) from turbine base to power system:

	Up to 100ft	200ft	300ft	400ft	600ft	800ft	1000ft
12V	8	6	4	2	0	2/0	4/0
24V	10	8	6	4	2	0	2/0
48V	12	10	8	6	4	2	0

Figure 18.4—Yaw bushing stuck to the grease on the tower top stub.

Figure 18.5—Alternator mounted on tower top stub, with power wires pulled through the hole in the top of the yaw bearing.

Install the alternator on the tower-top stub

Before installing the alternator you need to pull the wires all the way up through the tower from the bottom, then through the tower top. Leave a good bit of wire sticking out the tower top. Grease the tower top liberally with axle grease. Then install the yaw bushing, which is a disc of self-lubricated bronze or plastic sized to fit on the tower stub and inside the yaw bearing, also with a 1 inch diameter hole in it. It comes with our completed wind turbine kits, and is available separately from us also. Put some grease on the tower top and "stick" the bushing on there (it should stick to the grease, Figure 18.4).

To guide the three tower wires through the hole in the tower top and bushing, use a stick or something similar, and poke the wires through the hole in the yaw bearing and yaw bushing. Stick it through these holes so it comes out the bottom of the wind turbine yaw bearing, and so you can tape the three wires to it. Then have one person pull the wires through the wind turbine while another person carefully slides the entire turbine onto the tower top (Figure 18.5). The objective here is to keep the wires from crimping up and binding inside the tower top.

After mounting the alternator, connect all three output wires to the output studs on the stator. You'll also need to provide a strain relief system at the hole in the top of the yaw bearing, so the power wires don't get ripped out from the weight of the tower-to-ground cable, or get abraded through the insulation until they short out. A Kellum-type cable grip works great, and is available at any electrical supplier. A Kellum is shown in Figure 18.6.

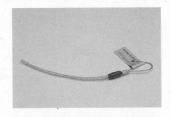

Figure 18.6—A Kellum-type grip strip strain relief for the power wires at the tower top.

Fit the tail

If it's a tilt up tower, raise the tower up so the wind turbine is about 5 feet off the ground. This will make it easy to fit the tail and the blades. The tail pivot is the stub that sticks up out of the alternator over which the tail fits. Grease the tail pivot with axle grease and then slip the tail boom over it.

Mount the rotor to the alternator

It's possible that you had to disassemble the rotor to transport it to the tower site. If so, assemble it now. Hopefully you followed the instructions in Chapter 16, *Blade assembly*, very carefully and labeled each blade and its position on the hub, so that everything goes back together exactly the way you assembled it the first time. Refer to the instructions in that chapter to make sure you get everything right. And, if the finish has dried out any since you first coated everything with linseed oil, now is a good time to give everything a final coat.

It's good practice to turn the stop switch on (short the alternator) so the wind cannot spin the blades while you install them. You made two 6 inch diameter steel hubs while working through Chapter 9, *Frame*, and one has a hole in the center so that it can fit over the grease cap on the alternator. The rotor will be clamped between these two steel hubs. Mount the hub with the hole on the alternator. Then fit the rotor to the alternator. You may have to use a rubber mallet (not a steel hammer!) to set the rotor down fully. Put the last steel hub on over the blades. Each stud should get one lock washer and one nut. Tighten the nuts down all the way. Rotate around to different bolts and keep tightening. The wood in the blades will "crush" slightly, and it takes a few times around to get things really tight. We don't use a torque wrench, but the bolts should be really, really tight! After the machine is about one month old, it's good practice to lower the tower and tighten these nuts again—they will loosen. Check them again in six months and at least once a year thereafter, during regular yearly turbine maintenance. Be sure to read Chapter 19, *Failures and prevention*, for more information about scheduled maintenance.

Figure 18.7—Balancing the blades with the turbine on the tower top.

Balance the blades

Figure 18.8—Placement of lead weights for balancing.

Once the rotor is fastened tightly to the alternator, all you have left to do is balance everything. We do this on the tower stub in lowered position, and it must be done on a day without much wind. To balance the blades you should have some wood screws, some lead weights, scissors (or tin snips) to cut the lead if needed, and a cordless screwdriver. We usually use "decoy weights"—these are the lead strips hunters use to weight down decoys (fake ducks and geese usually), and they are available at sporting goods stores. Otherwise, lead flashing or sheet of any kind is fine. Sometimes when we can't find lead we use steel bar stock. That works too, it's just larger for any given weight.

First disengage the stop switch (so that the alternator output wires are *not* shorted together) so that the alternator turns freely. The blades will find a resting point such that the heaviest part comes down to the 6 o'clock position. Take note of the heavy part and raise it up to 3 o'clock, then use one wood screw and lightly attach a weight at 9 o'clock. The weights should be located at or near the hub. Do not add weight too far out on the blades. We never put weight on the carved parts of the blades, only the backs or edges near the hub (Figure 18.8). If the weight you put on is not heavy enough you can move it out further, or add more weight as needed. If it's not a windy day this should go pretty quickly and easily. Sometimes different weights are required in different places—you just need to keep working with it until the blades no longer seem to have a a heavy side or a "preferred" position.

Raise the wind turbine

At this point you should have it all together! Before raising the turbine, it's good practice to run through a final checklist:

• System wiring to controller and battery bank complete, breaker turned on.

• Shutdown switch in on (shorted) position to prevent blades from turning.

• 3-phase plug at tower base is plugged in.

• All 3 tower output wires connected to stator.

• Yaw bearing, tail bearing, and main shaft bearing greased.

• Stator is centered between the magnet rotors and doesn't rub when rotors turn.

• All nuts and bolts on turbine and rotor tightened.

• All tower hardware double checked (guy wire clamps, turnbuckles, etc).

• 2–3 human helpers rounded up.

If everything looks tight, properly adjusted, and balanced, it's time to clear people and dogs out of the "fall zone"—that's anywhere in the circle around the tower base at a radius of the towers height. If you have guests that are not actively helping, keep them far away from the fall zone with their cameras—you might even considering marking the fall zone on the ground with spray paint or plastic tape if you have lots of guests. This is no time to be distracted, by anyone. Hopefully, you already know exactly how this procedure goes because you've first observed an expert raise a tower, and then test raised your own and adjusted the guy wires. So this procedure should go like clockwork for you and your crew, because you've already done it before in test raising your tower!

Figure 18.9—Preparing to raise the turbine. It's installed on the tower top, all the wiring is complete, the final checklist has been completed, and all spectators are outside the fall zone.

Slowly and gently raise the tower and turbine. Check all guy wires, turnbuckles, etc. and tighten if needed. Turn the turbine on with the shutdown switch, and hope for some breeze! If you've done everything correctly it should start up in very low winds and be producing usable power in 6 or 7 mph winds. The blades should be almost completely silent, however the alternator will usually make a "humming" sound once current starts to flow.

One last thing—be sure to "mouse" your turnbuckles! Simply thread a piece of copper wire through and around the turnbuckle body and threads so that vibration cannot slowly loosen them over time.

Figure 18.10—Almost up!

Energy non-conservation:

a true story!

The winter of 2006-2007 was very unusual way up here in the Northern Colorado rocky mountains. On December 22, 2006, we were hit with over 4 feet of snow in 24 hours. When the sun came out again only a day later, everybody up here in our community was doing just fine for power—cold temperatures and highly reflective snow cover make for excellent PV power output. But then the wind came up...

Our good friend and neighbor Tom H flies the 10 foot wind turbine design covered in this book. His battery bank was full from his PV panels. When the power started coming in from the wind, he needed to shut the wind turbine down. But with over 4 feet of snow on the ground, shoveling a path to the bottom of the tower where his shutdown switch is located was a long-term proposition; at least one or two days of shoveling. He had no automatic dump load controller to divert extra power from the battery bank, and his only controller was "manual"—use more power to offset what's coming in from the wind.

So, Tom did the exact opposite of what every reader of this book should do—he went around the house and replaced every single energy-saving compact fluorescent (CF) bulb with a power-hogging and obsolete incandescant, of as much wattage as possible. These had been stashed in a desk drawer for years, from before he installed a wind turbine. The combination of this tactic, plus leaving the television and stereo and every light in the house on 24/7, made for just barely enough load to compensate for what was coming in from the wind and solar.

Three days later, Tom was finally able to shovel a path to the wind turbine tower and throw the shutdown switch. Over the next few days, he went back through the house and replaced all the power-hogging incandescents with their proper CF replacements, and everything was back to normal.

19. Failures and prevention

Wind turbines have a hard life. Unlike automobiles or airplanes or almost any other machine, they are continuously exposed to very harsh conditions and usually get minimal attention until something goes wrong.

Like all machines with moving parts, wind turbines need occasional maintenance and repair or they will break down from time to time. Bearings wear out, metal fatigues and corrodes, wood erodes and rots over time, extreme weather events happen...the list goes on. Oftentimes the problems one has to deal with depend a great deal on the climate. A turbine that lives in a hot, dry and dusty climate will face a whole different set of problems than a machine that lives in a humid, salty marine environment.

Most failures stem from an "event cascade." That's when one tiny little problem that's not fixed immediately starts causing other problems, which all get worse and worse—until they all together cause a total failure of the turbine.

Attention to detail

Most failures we've seen could've been prevented by greater attention to detail during assembly and installation. Is the stator perfectly centered between the magnet rotors? Is all the hardware tight and did you remember lock washers? Are the blades well balanced? The list goes on. There are a lot of adjustments, fasteners, and electrical connections in a wind power system. A failure of any one could easily lead to much bigger problems and down time. For a wind turbine, Murphy's Law is direct and brutal—if it *can* break, it *will*. The weather won't cut your turbine any slack whatsoever. Very careful assembly and installation will save you lots of grief! An example is shown in Figure 19.1. This early wind turbine never got to make a single

Figure 19.1—Oops! Somebody forgot to tighten the cable clamps on a guy wire before raising.

watt because during the frenzied preparations for raising it, someone forgot to tighten the cable clamps on one of the side guy wires. The turbine got about 3/4 of the way up, then tipped to the side as the guy wire loop pulled loose from the anchor. Fortunately, nobody was in the "fall zone," so no humans or dogs were injured.

Proper maintenance

Many problems can be prevented by regular maintenance and inspection. We suggest that at least once a year, you lower and then carefully inspect the machine. Double check that all the hardware is tight. Inspect the condition of the wooden parts and refinish them if necessary. Check for smoothness in the bearing. It's a simple matter to disassemble the alternator from time to time and inspect and/or re-grease the bearings if necessary. A little bit of time spent on maintenance and inspection can save a lot of down time and possibly prevent catastrophic failure. Our recommended maintenance schedule comes later in this chapter.

Eyes, ears and nose

Use them! Does something look strange? Is the turbine making a funny sound? Is there an odd smell coming from the rectifiers? Even though the machine may still seem to be producing good power, any strange new behavior mandates shutting down the machine, lowering it, and inspecting it carefully. A loose or worn out bearing might cause the magnet rotors to collide with the stator on occasion, which is very audible. A partially burned out rectifier can lead to overspeeding and a stator burnout, or worse. Both are easy fixes if you catch them early. Like you would with any other machine, pay close attention to the wind turbine, if anything seems unusual—shut it down and keep it that way until you can carefully inspect the machine and resolve the problem.

Figure 19.2—Alternator hardware has vibrated loose.

Figure 19.2 shows Matt B's 15 foot wind turbine alternator. He noticed a strange clicking and clanking noise, it seemed like the stator was moving around, and so he shut the machine down. Then he noticed some nuts and bolts from the alternator lying on the ground at the tower base. Due to a slight lack of attention to detail (and thread locking compound) on the alternator nuts and bolts (not tightened enough), much of the hardware that held the stator in place had vibrated completely loose and fallen off! No damage was done. He caught the problem in time because something didn't sound right and didn't look right.

Some failure examples

In this section, we'll cover some of the different wind turbine components that have failed for us and our neighbors. Most could have been prevented by proper inspection and maintenance, and most were the result of an event cascade.

BLADE FAILURES

Very few blades fail by themselves if they are reasonably well designed, and built from appropriate materials. The most likely cause for a blade failure on a well built and well designed machine is overspeeding. If the blades run far too fast due to partial or no load on the alternator, they are at risk. The forces trying to yank the blades apart are related to the square of rpm (twice the rpm, 4 times the force). We have had very few blade failures,

Figure 19.3—Blade root failure at glue joints.

though. One was on our 20-foot diameter wind turbine, where we built the blades (each 10 feet long) from 8-foot lumber. The blades were made from laminated material and we had put several joints in each blade near the root of the blade. They all failed in winds over 80 mph (Figure 19.3). In retrospect this was a very bad blade design, and the failure was preventable.

If the blades are reasonably well made from good quality wood and the hub is strong, blade failure is unlikely. Most of the blade failures we have seen were the symptom of another problem. We have seen blades fail when they struck the tower, which could've been prevented had we designed the machine so that the blades had a bit more clearance. We've had failures when magnets were not well secured in the magnet rotors, then flew out and struck the blades. Overall though, the blades have not been

Figure 19.4—Expensive kindling!

problematic for us. One issue that does come up after time is blade erosion. Eventually dust will have its way with the leading edge of the blades. Unless the blades are regularly inspected and refinished as needed, or protected with leading edge tape (some older machines used copper flashing stapled around the leading edge) the blades will eventually erode.

MECHANICAL FAILURES

Mechanical failures happen on all machines with moving parts. Sooner or later the main bearing will fail, just like with the trailer that the bearing we use is designed for. You can increase bearing life by making sure it's always well lubricated and properly adjusted. It probably never hurts to always keep a spare bearing set on hand, they are quite inexpensive. Other mechanical failures are usually due to improperly tightened hardware, or neglect. All the hardware that's against wood will likely loosen over time as the wood crushes. The stator is always subject to vibration whenever the wind turbine is producing power, and the hardware that supports it could come loose over time. A minor mechanical problem is easily fixed, and a failure will usually lead to worse problems. Inspect the machine at least once a year and make sure all the hardware and the bearing are in good shape. For the first year, a turbine should be inspected more frequently and watched more carefully.

Another bearing that can wear out and fail is the yaw bearing on the tower top. When you installed the machine on the tower top stub, you should have inserted a flat, nylon or oil-impregnated bronze bushing between the stub and the yaw bearing on the frame. If you didn't, or if the bushing has worn out, the stub can wear right through the top of the yaw bearing, and the turbine will drop down onto the stub. So, this yaw bearing should be checked and greased regularly.

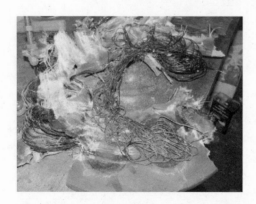

Figure 19.5—The sorry remains of a burnt-out stator.

ALTERNATOR FAILURES

Of all the problems we've seen, alternator failures are probably the most frequent (Figure 19.5). Alternators usually fail because the stator overheats. There are two common reasons for this:

• Too much power: The formula to figure heat dissipated in the stator, in watts, is $P = I^2 \times R$. Consider the stator as we build it for the 10-foot diameter wind turbine: The coils occupy a ring of approximately 6 inches inner diameter and 14 inches outer diameter. Calculating the area of this ring on both sides of the stator, we believe that this type of stator, cast in resin, can dissipate no more than 3 watts per square inch, and in a hot environment probably less. If you know the wire gauge you used to wind your stator and the weight of the wire, you can refer to a wire gauge chart and calculate the resistance of one phase in the stator. Roughly speaking, you can then calculate how much heat the stator will have to dissipate at any level of output. If you see sustained output over this "limit," you either need a more powerful alternator or you need to adjust the machine to furl in lower winds. See Chapter 20, *Scaling it up and down*, for more information on stator resistance and heat.

The easiest way to compensate for too much power coming in is to figure out a way to make the tail weigh less, so the machine furls earlier. Another thing to check if the machine is not furling in low enough winds is the air gap between the magnet rotors. If it's too wide and the cut-in speed is too high, the blades will run faster and it will likely produce more power and furl late. For the 10-foot wind turbine detailed in this book, you should not see sustained output over 600 – 700 watts, unless you build with the larger round magnets and appropriate stator (Chapter 12, *Stator*), in which case sustained output of 900 – 1,000 watts is reasonable.

• *Bad or broken electrical connection(s):* These can easily lead to alternator failure and other problems. If just one wire between the turbine and rectifier breaks or becomes disconnected, then the three-phase alternator is suddenly running single phase! This usually causes the blades to run at higher rpm, which will usually delay furling, and in many cases the machine will still produce near full power. So, all the power the wind turbine is producing is then running through a single phase instead of three. This can often overheat one phase of the alternator and lead to stator failure. Another issue with running single-phase (and an audible symptom) is severe vibration. We had one machine that was quite reliable for a couple years and then a connection came loose at the base of the tower. Within one hour, all the hardware that had reliably supported the stator for 2 years rattled loose and the stator was floating freely between the magnet rotors. Fortunately, we heard the sound and shut the machine down—it was an easy fix. A partially blown three-phase rectifier can have exactly the same effect, and a broken connection can also lead to a rectifier failure! Everything is interconnected.

RECTIFIER FAILURES

As we just mentioned, a rectifier failure can easily lead to alternator failure, overspeed, possible blade failure, and worse (OK, it doesn't usually get much worse than those three!). Usually rectifiers fail when they get too hot (a heat sink that's too small) or when they're overloaded. It's good practice to use a large heat sink and pay attention to it for a while. If it gets any more than warm to the touch, consider a larger heat sink. Use rectifiers that are rated for at least 50 percent more current than you ever expect to see from the wind turbine. Sometimes no matter how careful you are, stuff happens. A close lightning strike can take out a rectifier, and if you're not on top of things that can lead to other failures.

THE EVENT CASCADE

Unfortunately, oftentimes a problem comes up and leads to other problems very quickly. Sometimes it's difficult to figure out what caused "the problem" in the first place! Consider the machine pictured in Figure 19.6. We believe it all started with a rectifier failure, because all the wiring was in good condition but one rectifier was blown. The rectifiers were only rated 30 amps, and it was a 48 volt system, so the rectifiers had no safety factor. It was a very windy day (gusting over 80 mph) and the owners had left the machine on and gone off to work. All of this could've been prevented if they'd just shut down the machine! When they got back from work, bits of the machine (tail, magnets, stator parts, blade parts) littered their yard.

Figure 19.6—Catastrophic wind turbine failure, caused by an event cascade that started with a single failed rectifier.

Nobody was there to watch, but here is what we think happened: A rectifier failed, forcing the alternator to run single-phase and allowing the machine to overspeed. It probably didn't furl normally due to overspeeding, and one phase of the stator had to do all the work, causing that phase to overheat. The insulation on the wire in the coils and the resin got hot (probably over 400 degrees F). The stator was most likely smoking hot, it swelled up and warped, and then came in contact with the magnet rotors. The friction between the fast spinning magnet rotors and the stator created a great deal of heat in the magnet rotors, overheating the magnets so that they lost their magnetism and melting the resin around them. As a result, the casting around the magnet rotors failed and about half the magnets flew out—the magnets themselves were "dead," having permanently lost their magnetism from friction overheating. Two of the blades were unfortunate enough to strike the flying magnets. The magnets felt like large caliber bullets to the blades, and the blades were chopped to pieces. Now, with big parts of two blades missing, the machine was suddenly running in very high winds very much out of balance, jumping around so much that the whole tail boom fell off. The owners are lucky it didn't take their whole tower down.

That was perhaps the ugliest failure we've ever seen, and the start of the event cascade was a part that cost only about $5. It could've likely been prevented with a bigger rectifier. It's very common that one problem will lead quickly to another, and that's a good reason to do everything you can to set it up right in the first place and pay close attention to your wind power system.

TOWER FAILURES

A wind turbine tower should cost at least as much as the turbine itself, but the tower commonly gets neglected in all the excitement of watching a new turbine spin. The two primary types of tower failures that we've seen are:

Figure 19.7—Ouch! The blades hit the upper guy wires, the turbine ran out of balance, and the entire unit hopped right off of the tower-top stub.

• *Blades hitting the tower:* This can happen when the sweep of the turbine blades is not correctly factored in to the design of the tower-top stub and the top set of guy wires. The force of the wind is tremendous on the blade tips, and in operation they can bend back toward the tower by a few inches. If the guy wires are attached to the tower too high, a deflecting blade can hit one. The reason the frame of our wind turbine is tilted back slightly is to avoid this problem, but the location of the top guy wire attachments is still critical. With a large chunk out of one or more blades, the rotor then runs in an extremely unbalanced condition. In Figure 19.7, that's exactly what happened to our neighbor Scotty's turbine! He reinforced the tower top with some extra guy wires, and the blades hit them. The entire machine hopped right off its pivot from the vibration of unbalanced blades, and the tail fell off as the turbine was hanging by its own output wires.

- *Tower support failure:* The stresses on a tilt-up tower are greatest when it's being raised or lowered. But that's when you are likely paying the most attention to details, and problems like loose cable clamps, improper guy wire length and failed turnbuckles are usually detected early. More insidious are problems that slowly happen over time, as the tower hardware vibrates, heats, and cools. Inspect your tower even more often than you inspect your turbine! The critical hardware is on the ground, after all, and you can inspect it frequently while taking a short walk around with a relaxing beverage. Check for anything that is loosening up, and tighten it immediately. Take precautions against such movement, too—many turbine installers actually thread a piece of sturdy wire through each turnbuckle and wrap it around, to prevent tower harmonic vibrations from slowly turning and lengthening the turnbuckle, thus loosening the guy wire.

One spectacular tower structure failure that happened to a neighbor of ours recently was caused simply by a lack of attention to detail. Tim S. was flying one of our experimental 17-foot turbines, and as usual installation day was chaos. He was short one turnbuckle, and so he instead used a nylon tow strap to secure the end of the gin pole to the concrete earth anchor. Everyone agreed, "You had better replace that with a proper turnbuckle within the next few days, or the nylon strap will abrade and fail." Six months later, the nylon strap did indeed wear through and fail, causing the entire tower to topple while the turbine was in operation. The aftermath is shown in Figure 19.8, but fortunately the only damage was a set of ruined blades—once again, very expensive and labor-intensive kindling. This lesson was rather harshly learned, plus an additional lesson: the gentleman with the digital still/movie camera who watched the turbine fall was too aghast at what was happening to point the camera and hit the "record" button!

Figure 19.8—A "temporary measure" during installation caused this grim tower collapse. The nylon strap securing the gin pole to the earth anchor should have been replaced by a real steel turnbuckle immediately.

Regular maintenance

As we've tried to pound into your head repeatedly throughout this book, wind turbines (even expensive commercial ones) are not something you can install and forget about! We recommend that you follow the maintenance schedule outlined below, or some reasonable facsimile thereof.

For the first month after you first erect the turbine:

• Shut it down while you are away, unless you are positive that winds will be gentle. After a month, you can leave it running while you are away (as long as you installed an automatic controller and dump load).

• Watch it regularly, including monitoring power output.

• Listen for strange sounds, sniff for strange smells, and watch for odd behavior.

• Check the pendant cable from the turbine to the ground weekly to see how much it has twisted, unplug it at the tower base, untwist the wires, and plug it back in. This will give you some indication of how frequently you'll need to check this in the future.

• One month after you install the turbine, lower the tower and inspect everything, including the guy wire anchors, turnbuckles and cable clamps. Check all bolts on the turbine itself, and re-tighten if needed.

Six months after you erect the turbine:

• Lower the tower (or climb a non-tilting tower), and firmly re-tighten all the bolts and nuts that hold together the blades and alternator.

• If you used linseed oil to finish the blades and tail, re-apply it in another thick coat or two.

• Check all guy wire anchors, turnbuckles and cable clamps.

Yearly (or every six months in harsh climates):

• Lower the tower, (or climb a non-tilting tower), and check all the bolts and nuts that hold together the blades and alternator. Re-tighten as needed.

• Apply more linseed oil finish to the blades, wooden hubs, and tail vane.

• Check all guy wire anchors, turnbuckles and cable clamps.

• Refresh the grease on the yaw bearing and tail pivot bearing. Grease the main bearing if needed. To do this, remove the cotter pin on the main bearing, re-move the main bearing nut, and apply axle grease liberally to the bearings. Put the nut back and get it resonably tight, then back it off until you can insert the cotter pin. Once the cotter pin is in, back the nut off as much as the cotter pin allows. You do not want this nut tight, that would make the alternator too hard to turn.

Always:

• Regularly check your pendant cable for twisting. If twisted, unplug it at the tower base when the wind is not blowing, untwist, and plug it back in. You should have an indication of how often this is needed by monitoring it closely for the first month the wind turbine is flying.

• Listen, look and sniff for strange changes that might indicate a problem starting to happen. If you catch and fix problems right away, no damage is usually done to the turbine. If you let the problems get worse, they can cause other problems—an event cascade that can lead to total failure.

• Monitor your power output—a big change in it could indicate a growing problem.

• Watch the turbine—if something seems to run wobbly or just doesn't seem "right," the turbine probably needs adjustment. You can even use binoculars to check for loose nuts and bolts and such.

• Enjoy the fact that you are making electricity for your power system using the free "fuel" of the wind!

ALTERNATOR ADJUSTMENTS

Most alternator adjustments can be made by re-positioning the stator between the magnet rotors. This might be needed if the stator mount loosens or gets out of whack, or if the magnet rotors start scraping against the stator. However, in rare cases you may need to remove the front magnet rotor to replace the stator—for example if you are upgrading your power system to a different voltage, or if the stator becomes damaged. It's a fairly simple operation, as you discovered in Chapter 14, *Alternator assembly*.

NOTABLE QUOTES

"Wind turbine mechanics are not like auto mechanics. They are few and far between. Best if you can fix it yourself."

Dan Bartmann, www.Otherpower.com

This procedure requires special tools called "jacking screws" to force the magnet rotors apart, which you probably fabricated for that chapter. Follow the same slow and steady procedure you used to assemble the alternator, in reverse, if you need to take it apart.

Troubleshooting

In this section we'll go over some of the symptoms of impending problems that we've seen in the past, and their possible causes and solutions. These things to watch, listen, and smell for apply to *all* small wind turbines, not just the design presented in this book. Once again, a warning and reminder—wind turbines are *not* "set and forget" units like solar panels. You should make a habit of paying attention to what your turbine is trying to tell you!

• I've erected the wind turbine and the breeze has come up, but it turns very slowly or not at all.

Did you turn the stop switch off (so that the 3 phase turbine output wires are *not* shorted together)? If the turbine is not being stopped by the switch, there is probably a short somewhere else in the system. The most likely place to find a short is in the three-phase rectifier. You can disconnect the rectifier (make sure the stop switch still works) and see if it spins freely then (don't let it spin fast though, and be sure to turn it back off with the kill switch quickly). If it turns then the problem is likely the rectifier. If it's still not turning you can check the resistance between each of the 3 leads with an Ohm meter. If one phase has significantly less resistance then you know you have a short somewhere between those leads—or (unlikely) a defective stator.

• The turbine seems to be running well, but it wobbles back and forth a bit (or the tower shakes).

Sounds like you need to balance the blades again. Blades can sometimes go out of balance by themselves with time—especially if they're not well finished and can absorb water.

• The blades spin really fast and make a whooshing sound and I'm not getting any power!

Yikes! Scary, shut it down immediately (turn the stop switch on)! Something is disconnected and your turbine is "freewheeling" with no load on it, a dangerous condition. Some possibilities are: the rectifier is disconnected, the rectifier has failed, or you've not got the rectifier hooked to the batteries. Another possibility, if you have a breaker or a fuse between the rectifier and the battery, it could have tripped or blown.

• **The blades spin kind of fast, make a whooshing sound and I'm not getting lots of power. The machine also vibrates more than normal.**

Well—if it has been doing this from the start you might not know what "normal" is, but if the problem comes along after time this is a sign that one of the three leads has become disconnected, or your three-phase rectifier is partially blown. If the rectifier fails, or one lead becomes disconnected, then the alternator is running single phase and the blades will overspeed. This will also likely cause the machine not to furl properly (it may wait for much higher winds before it furls). In this condition, part of the stator is doing all the work and it will likely overheat and or burn out in higher winds, so the problem must be addressed. You can check continuity in the lines with an Ohm meter. If you have continuity and similar resistance in all 3 phases then the problem is almost surely a blown rectifier.

• **It makes a scraping sound and doesn't start up well in low winds.**

Sounds like the magnet rotors are hitting the stator. You should shut down the machine until this is fixed. The stator might have been out of adjustment (not centered between the rotors) when you erected the machine. It's possible that some of the hardware holding the stator on has come loose or fallen out. The roller bearings might be worn out, or need adjustment. If the bearings get loose, then the rotors could hit the stator. Worst case scenario is that your stator overheated and warped. If this is the case and the stator is still working and making power, you might be able to re-adjust it—but odds are you'll need to make a new stator. If the stator overheated you need to take measures to prevent that in the future. A lighter tail will help the machine to furl in lower winds and help prevent stator overheating.

A stator could also overheat for other reasons—an improper load (wrong battery voltage or hooking the wind turbine directly to heaters) could cause it. A burned out rectifier (previously discussed) or a disconnected lead that allows the machine to run single phase could also overheat the stator. If this sort of situation is allowed to go on too long, the magnets will rub on the stator and the friction may heat the magnets to the point where they lose some of their magnetism. If this happens a new stator and new magnet rotors are required. This sort of thing is rare—but it is yet another good reason to actually watch and listen to your wind turbine regularly.

• It's been cold and we've had a bit of an ice storm. Ice has built up in the alternator and the machine has seized.

This is rare. But sometimes if it gets cold and there is no wind (the machine is stopped for a while) and you have freezing rain or fog, the alternator can get packed up with ice and not run. Large utility-scale turbines actually have icing sensors that shut the machine down when ice builds up. You either have to wait it out, or melt it off. Here in Colorado, the sun usually comes out quickly after a storm to melt the ice, so we usually wait it out. If you can't wait or if the ice storm is prolonged, a good way to do this (and you could build this feature into your rectifier box) is to send full battery voltage back up the line to the alternator (bypass the rectifiers). About 30 – 60 seconds of this will warm up the stator and start to melt the ice. It also has the effect of making the machine want to turn. You may be able to melt the ice by just sending battery voltage up only two leads for about 30 seconds. If that doesn't do it, try sending it up another phase. Don't leave power applied to any one phase for too long (you don't want to overheat the stator) but be persistent.

Our experience suggests that this is usually quick and easy. We've only had this problem a couple times in many years—it may be more frequent in less windy, more humid environments. It's really much easier to wait it out and let the sun melt the ice. Be extremely careful if your machine is iced up, and the wind starts to blow and turn the blades! This can be a dangerous condition. Chunks of ice could fly off the turbine at high velocity and travel a long distance, and/or the machine could try to spin up and run out-of-balance. If you see these conditions develop, shut the machine down with the stop switch.

• The turbine has been making squeaky noises like an irritated hamster, and is not turning as well into the wind as it should.

Hopefully you put a bushing between the tower top and the turbine itself. If it's making noise or is having difficulty yawing, you should lower the tower, check the bushing and apply more grease to the tower stub. Get on top of this problem sooner rather than later—if you don't, the top of the wind turbine frame will eventually wear through and the entire turbine will drop down through the yaw bearing.

• I really don't have time for all this—I think I'd rather just go fishing.

Sorry! Wind turbines take quite a bit more work to install and maintain than do solar panels that you can just set and forget. What were you thinking?

• **The alternator makes a humming sound but it seems to be rather loud and it rattles. Everything else, including the output, seems normal.**

Some tower designs may amplify the sound of the alternator, especially if you use thin walled tubing or have loose couplers between pipe and tubing sections. Another possible amplifier for the alternator is the tail vane. If the tail vane is loose it may act as a "sound board" for the alternator. Check to make sure the tail vane is bolted on tightly, and if you want further insurance, you can apply caulk between the tail vane and all the steel pipe and bar stock that it comes in contact with.

• **In low winds everything seems quite good, but in higher winds the machine tends not to speed up much and it's not producing as much power as I'd hoped.**

Sounds like the blades are stalling. Low battery voltage is one possible cause. On the machines where we use the more efficient heavy duty alternator, the blades are inclined to stall in higher winds unless there is a fair bit of resistance between the alternator and the wind turbine. This resistance could simply be a long (or relatively thin) line between the turbine and the batteries—so these machines are more suitable for placing some distance from the battery bank. Or, you could add a resistor between the rectifier and the batteries. See the sidebar *"A more powerful alternator"* in Chapter 12, *Stator*, for more information on adding resistance—and also Chapter 20, *Scaling it up and down*.

DOG HAIKU
Look, listen and smell
Is your turbine now failing?
It tried to tell you

20. Scaling it up and down

We get many, many questions regarding scaling this 10-foot wind turbine design up to make a larger machine, and down to make a smaller one. It seems like one should be able to model everything with a spreadsheet and have it spit out a new design, right? Well, yes and no. You can certainly plug in the numbers for an alternator design, after you calculate a power curve for rpm versus horsepower coming out of your blades. If the alternator is of a reasonable design, there are formulas that could get you in the ball park. But there are many variables which are difficult (at least for us) to predict. Some of these include the size and shape of the coils, the space

Figure 20.1—Our scaled-up, 20-foot diameter wind turbine. It's the biggest we've built, and is still experimental.

between the magnets, and the air gap between the magnet rotors. Mechanical clearances also change when scaling up or down, with larger machines needing greater distance between moving parts. Variations in the basic design and geometry of an alternator, along with the tolerances held during the construction, make it very hard to accurately design for the power curve in advance. Since matching the power curve of the blades to the alternator can be tricky stuff (see Chapter 4, *Electricity from a spinning shaft*), we tend to get close using empirical data, and extrapolate information from previous machines we've built.

When designing a new machine, we usually start by considering the amount of magnetic material needed based on the swept area of the rotor, then build appropriate magnet rotors. Next we wind a single test coil, and bench test it for voltage versus rpm between the magnet rotors. From there, the math for determining the number of turns and wire gauge is linear and easy.

In this chapter we'll discuss some of the things to keep in mind if you want to build a similar machine of different size, and we'll touch on both larger and smaller machines that we've built. Figure 20.1 shows George in front of our experimental 20-foot diameter turbine. There's only one of these flying so far, so the designs we'll talk about in this chapter are the 7-footer and the 17-footer. These are experimental too—but we have had enough of them flying for enough time to present some details for the adventurous enthusiast.

Power in the wind, again

Yes, you've seen this material before! And it's all critical to scaling a wind turbine design up or down. The power you can get from a wind turbine is related to two things:

• The cube of the wind speed. If you increase wind speed a little bit, you have a lot more power available. The only way to increase wind speed is a taller tower.

• Swept area, which is the same as the square of the diameter. In other words, if you double the blade diameter you have 4 times the swept area and 4 times the power.

Since the tip speed ratio of the wind turbine always remains about the same, it's safe to assume that rpm is related inversely to blade diameter. Let's consider what happens if we double blade diameter:

• Double the blade diameter, and the blade tips move at the same speed, but the rpm will be half—because when you double the diameter of a circle, the circumference is also doubled.

• The cut-in speed will also be about half. So if our 10-foot diameter machine cuts in at 140 rpm, we would design a 20-foot diameter machine to cut in at around 70 rpm.

• The swept area and the power available from 20-foot blades is 4 times that of 10-foot blades. So the alternator needs to produce 4 times the power, at half the rpm. Roughly speaking, this means you need about 8 times more alternator if you double blade diameter.

We have actually built a couple of 20-foot diameter wind turbines here, and these are some statistics to ponder: The 10-foot turbine described here in this book has 12 inch diameter magnet rotors, a total of 24 cubic inches of magnetic material, and about 6 pounds of copper in it. Our 20-foot diameter wind turbine has 22 inch diameter magnet rotors, about 135 cubic inches of magnetic material, and about 25 pounds of copper in the stator. The 10-foot diameter turbine produces about 700 watts at around 400 rpm, the 20-foot diameter machine produces about 4 kw at 200 rpm. The actual figures depend on line loss, battery voltage, etc.

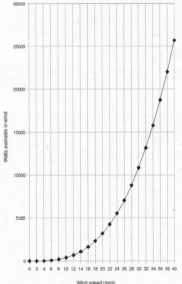

Figure 20.2—The exponential curve of power available in the wind to a 10-foot diameter wind turbine.

Matching the alternator and blades

Designing a new wind turbine boils down to a basic problem: Power available in the wind increases exponentially with the cube of the wind speed (Figure 20.2), but power produced by an alternator increases only linearly with the rpm. Fortunately—thanks to the furling system—we don't have to deal with any increased power from winds over 28 mph, and can focus only on winds between the cut-in speed (7 mph) and furled position (28 mph). The ideal alternator would track this curve exactly in those winds. Consider yourself warned that if you undertake a new wind turbine design of your own, you'll be in for some trial and error building and testing of alternator designs!

DESIGNING A NEW ALTERNATOR

Here's how we go about it at Otherpower.com HQ:

• Estimate the amount of magnetic material needed in the alternator based on the swept area of the new turbine. We simply scale this figure up or down from what we used in previous successful and well-tested designs. Our general rule of thumb is that the thickness of each magnet should be about the same as the thickness of the stator itself. In the example above, the 10-foot diameter turbine from this book uses 24 cubic inches of NdFeB magnets for a swept area of 78.5 square feet, and our big scaled-up 20-foot turbine uses 135 cubic inches of magnet for a swept area of 314.2 square feet. The scaling for magnetic material is not exact, since magnet blocks are available in only a small variety of dimension options. Custom sizes are difficult and expensive to obtain, and rare-earth magnets are nearly impossible to cut or machine.

• Decide on the diameter of the magnet rotors and number of magnetic poles. Again, this is also partially determined by the dimensions of the magnet blocks you have available. The closer together you place each magnet to its neighbors on the rotor, the more magnetic flux potential is wasted by leakage between the magnets instead of being directed (by the complete magnetic circuit) through the coils in the stator. Spacing the magnets very closely does increase flux through the coils, due to the increased amount of magnetic material—but you don't get as much flux per unit mass that you add. Closely spaced magnets give you less flux for your magnet dollar. A good compromise is to space the magnets about one magnet-width apart at the closest dimension when using rectangular magnets, or about 1/4 to 1/2 the magnet diameter when using round ones. Keep in mind also that to easily make three-phase AC output, you need to maintain the ratio of 4 magnetic poles for every 3 coils. This also affects the final size of your magnet rotors.

• Calculate the desired cut-in rpm. The turbine starts spinning at start-up wind speed, and by the time it reaches cut-in wind speed for the alternator we want it spinning near the designed TSR. Here is the math that we used to calculate the desired cut-in rpm for the 10-foot wind turbine in this book, for a TSR of 6 and cut-in wind speed of 7 mph:

Blade tip speed at TSR of 6 in a 7 mph wind = 6 × 7 = 42 mph;

Distance travelled by blade tips in one rotation = pi × diameter = 31.4 feet;

(42 miles / 1 hour) × (5280 feet / 1 mile) × (1 hour / 60 minutes) × (1 rotation / 31.4 feet)

= 118 rpm

Great! Simply design the alternator to cut in at 118 rpm where the turbine will be running at TSR of 6 in 7 mph winds, and everything works fine, right? Wrong. We were fooled by this a few times until we discovered what was actually going on. Our early machines stalled badly at cut-in wind speed, though the math above was accurate. What happens is that before cut-in, the rotor is actually overspeeding a bit and running above the design TSR, since there is no load on the alternator. To prevent this problem, raise the calculated cut-in rpm for your design by 20 to 30 percent to allow the blades to gather some momentum before the alternator cuts in. For our 118 rpm calculated cut-in, we settled on an actual cut-in of 140 rpm. If you are designing a new turbine this is one place where trial and error will be essential, as all the very small details of your blade and alternator design will affect it.

• Build the magnet rotors, and determine the exact size and shape that the coils should be, based on the size and spacing of the magnets on the rotors. The hole in the center of each coil should be about the size of one magnet. Wind a single test coil to the exact size and shape needed out of any gauge of magnet wire, and keep careful records. The test coil can just be a gross estimation from previous designs as to how many coils of what gauge wire to use—the only requirement is that it fits tightly in the space you have available for each coil on the stator. You'll determine the proper wire gauge to buy in the next step.

• Fasten the test coil between the magnet rotors so that the hole in the coil is centered over the path of the moving magnets (Figure 20.3). Spin the magnet rotors either by hand with a helper, or with an electric motor such as a drill press or lathe. Measure both rpm and AC voltage from the coil. A laser tachometer is a great tool for measuring rpm, though you can get fairly accurate results with a clock, turning the rotors exactly once or twice per second. The exact rpm at

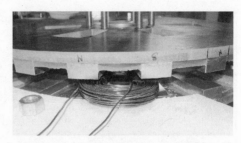

Figure 20.3—Using a single test coil to determine the number of turns of what gauge wire.

which you turn the rotors is not important, since voltage is directly related to rpm—but it *is* important that you record the voltage versus rpm readings from the test coil.

The number on your meter is an RMS (root-mean-square) measurement if you have a "true RMS" meter, and a close approximation of that if you have a regular meter. See the sidebar, next page, for more information on the vagaries of metering. Now, multiply by 1.4 to give the peak AC voltage, since that is what you'll see when you rectify the AC into DC. Multiply that by the number of coils in each phase to get the voltage of one phase—normally there will be 1 coil in each phase per every 4 magnets on 1 rotor, so the 10 foot turbine has 3 coils in each phase. The alternator is normally wired in star (see Chapter 4, *Electricity from a spinning shaft*), so multiply that figure (the voltage from one phase) by 1.73 (the square root of 3). This gives you the final peak AC voltage you would get from the alternator if you built all the coils like your test coil. Remember also that the actual DC voltage from the alternator at any rpm will be 1.4 volts *less* than what you've calculated and tested here, because of the voltage drop from the rectifiers.

Unless you are extremely lucky or know a whole lot more about math and computer modelling than we do, the test coil will be completely wrong! But now everything is linear—double the rpm, and voltage doubles. Double the number of turns in each coil and fit it into the same space by decreasing the wire size by half (3 AWG sizes) and voltage doubles. You can then use some simple algebra to calculate how many turns of what size wire to use in your stator. Always aim for a minimum of resistance in your coils—use the thickest wire that will fit, or multiple strands in hand.

Meter madness

Two plus two equals four. And 120 volts AC equals 120 volts AC, right? Wrong—it all depends on the meter you use, the AC waveform your are measuring, and how your meter makes its calculations. If you are measuring the AC voltage of a pure sine wave, any inexpensive multimeter will do the job. So grid-powered household electrical outlets and the output of most gasoline generators will read out just fine, 120 volts. But what if the waveform is not a sine wave? That's when things start to get weird.

Most inexpensive inverters put out what's often called a "modified sine wave," giving a waveform that looks like stair steps. It should really be called a "modified square wave," and is shown in the drawing below. If you measure this with a plain old inexpensive meter, 120 volts AC will actually read somewhere around 95 – 105 volts. To get the right reading, you need a more expensive "True RMS" multimeter. RMS stands for "root-mean-square," and is basically a sort of exponential averaging—the name comes from the fact that an RMS measurement is the square root of the mean of the squares of the meter readings as you take them.

What if the cheap multimeter you bought for 10 bucks is not true RMS? Don't worry about it. Just be aware that your readings for voltage, amperage and therefore power might be a bit off when measuring odd waveforms, like those from a modified sine wave inverter. If your alternator design is reasonable, you can assume that the waveform will be a pretty decent sine wave and your reading will be quite close to RMS. And for the purposes of your test coil and new alternator design here, peak voltage is what you are most interested in anyway, since you'll be rectifying to DC from AC. The peak voltage of typical 120 VAC household power from the utility is about 160 volts. So, you can multiply your reading by 1.4 (roughly the square root of 2) to get peak.

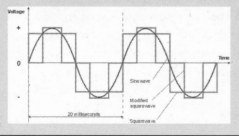

Now you have a rough estimate of how to wind your stator to get the right cut-in speed for your blades. The next consideration is how the overall power curve of the alternator matches the overall power curve of the blades.

RESISTANCE

To start with, consider the resistance of the stator. If you are designing a wind turbine of (for example) 20 feet in diameter and you expect the output to be 4 kw, you can calculate how efficient the stator will be at that level of output, and therefore how much power will be wasted as heat in your stator. An easy way to determine this is to weigh the test coil and refer to the wire gauge chart in Appendix B. Assuming the actual coils are each of equal weight, all you need to know is the wire gauge of the final coils. Measure the weight of the test coil, then calculate the the length of the wire you used in one coil from the chart to find the resistance of each coil. It's not really feasible to actually measure such low resistances with a multimeter; when the resistance gets high enough to measure, it's already far too high for your stator.

In three-phase alternators, you can figure the resistance of the stator will be twice that of a single phase—wired in star, the resistance of 1 phase will be roughly that of 2/3 of the coils in series. For the 10-foot turbine featured in this book, the final resistance is that of 6 coils in series.

Power wasted in the system (and the stator is part of the system) is calculated by:

$$P = I^2 \times R$$

Power (wasted) = Amps² × Ohms

There are two rules with stator resistance that you do not want to break with your design, or *it* will likely break:

• Don't get below 50 percent efficiency. For example: if you are planning a 4 kw, 20-foot turbine and the test coil math you did above shows it to be less than 50 percent efficient at maximum output level, you are likely in trouble—the stator has a high chance of burning out. Since the magnet rotors are already built, it makes sense to consider scaling the machine down a bit, perhaps to a 17-foot diameter 3 kw machine. Plan to wind the coils with somewhat thicker wire and a few less turns to give a slightly higher cut-in rpm. This also gives you more of a safety factor in high winds.

• Consider the heat dissipation capability at maximum output for the stator in watts per square inch, in only the area of stator directly over the coils on both sides. If the stator is required to waste more than about 3 watts per square inch in that area, you are likely in trouble again from the stator overheating and then warping. We try to keep wasted heat far below that level just by how we design a new stator—but the other way to deal with it is to make the machine furl earlier so it never sees those higher sustained outputs (*Chapter 5, Furling and regulation*).

Your actual results may vary quite a bit, and it's very likely that your first stator will need some adjustment. Look at the straight line above the wind turbine output curve in Figure 20.4. That's the output of an alternator that's too powerful for its rotor, and the machine will stall badly, hovering right around cut-in speed and making little power. It is also how your output will look if you built the more powerful alternator detailed in the sidebar in Chapter 12, *Stator*. In drastic cases of mis-match, you may have to build a new stator. But if the mis-match is small (like the one shown), there are two things you can try:

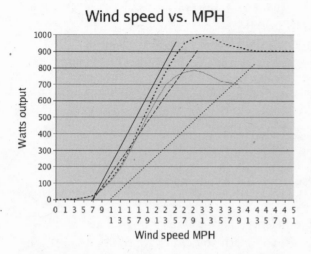

Wind speed vs. MPH

Watts output

1000
900
800
700
600
500
400
300
200
100
0

Wind speed MPH
0 1 3 5 7 9 1 1 1 1 1 2 2 2 2 2 3 3 3 3 3 4 4 4 4 4 5
    1 3 5 7 9 1 3 5 7 9 1 3 5 7 9 1 3 5 7 9 1

Figure 20.4—The solid, straight line above the wind power curve shows the output of an alternator that is too powerful for the blade set. The dotted, straight line under the curve shows the effects of opening the air gap too far. The straight, dashed line shows the effect of adding resistance to the line. The two turbine performance curve lines show the output of the normal stator (lower line) and the more powerful stator (upper line) detailed in Chapter 12, *Stator*.

• Increase the air gap. A small increase or decrease of the distance between the magnets in the two magnets rotors will make a large difference in the output. This method can be used to a small extent, but major changes can cause problems. Look at the lower dotted line below the wind power curve in Figure 20.4—that's the effect of opening the air gap. Notice however that it also raises the cut-in speed and flattens the power curve overall. Opening the air gap will allow the blades to run faster over the full range of wind speeds, which could be good or bad. If they run too fast in high winds, the stator could burn out. If the cut-in speed is already too high, you can close the airgap to a point, but you do need to keep good mechanical clearance between the magnet rotors and the stator to avoid rubbing.

There is one other adjustment you can make:

• Add resistance to the line. Look at the dashed line below the wind curve in Figure 20.4—it shows the effect of adding resistance to the system. Adding resistance will flatten the power curve without changing the cut-in speed. So, if the alternator is bit too powerful for the blades but the cut-in speed is good, adding some resistance to the line can help match the overall power curve of the alternator to the blades. We give details on how much resistance to add in the *More powerful alternator* sidebar in Chapter 12, *Stator*. The alternator doesn't care where the resistance is located, it can be in the coils themselves or in the line from the alternator to the battery bank. It's much easier to add resistance on the ground than to wire and cast a new stator!

A smaller and a bigger turbine

At the time of writing this, we have built three of the smaller 7-foot diameter wind turbines and six of the larger 17-foot diameter wind turbines. So far these machines seem to be running well and we have not seen serious problems, but they have been tested less and there are far fewer in the field than our 10-foot machines, which are described in detail in this book. It takes a lot of testing and a lot of machines in the field to really work the bugs out enough that one can feel confident about a new wind turbine design. So, we still consider these machines to be "experimental" and have not provided detailed plans. Rather, we provide the basic details and dimensions of the construction in hopes that should the reader dare to build one, they can figure it out based on the plans for the 10-foot diameter machine with the changes described here.

7-foot diameter, 300 watt wind turbine

So far this has proven to be a nice little wind turbine. It won't likely make much of a dent in the energy requirements of a normal home, but for a small off-grid cabin or any situation where power requirements are not too high, it can do a lot! It sweeps about half the area as the 10-foot turbine and therefore we would expect about half the energy. In our own testing we often see sustained output of over 400 watts with occasional peaks over 800 watts.

Figure 20.5—Our new 7-foot turbine design.

We especially like this design for 12 volt systems. It's much easier (and less expensive) to handle the current from the smaller machine at 12 volts. We feel that the output of the 10-foot turbine is pushing the limits for a 12 volt system.

The 7-foot diameter machine (Figure 20.5) is very similar to the 10-foot design detailed earlier in the book. All of the metal work is practically identical with a couple of important exceptions. The alternator is only offset 4 inches to the side of yaw bearing, instead of the 5 inches called for in the 10-foot plans. The tail boom is only 4 feet long instead of 5 feet as called for in the 10-foot diameter wind turbine plans. The tail bracket (the flat bar stock that

Figure 20.6—Aluminum magnet layout template for the round magnets used in the 7-foot machine.

Figure 20.7—The single magnet rotor in place.

holds the wooden tail vane) is also shorter, it's 36 inches long. Otherwise though, the 7-foot machine is nearly identical to the 10-footer. We use the same hub, spindle and bearings, and the hardware list is the same as is the basic assembly procedure.

MAGNET ROTOR

In Chapter 12, *Stator*, we mention the possibility of using larger 2 inch diameter x 1/2 inch thick disk magnets to build a more powerful alternator. This 7-foot turbine uses these larger magnets, but it only requires 12 of them and it only uses a single magnet rotor! Many who understand alternators and magnetic circuits will likely complain that using a single magnet rotor does not involve a complete magnetic circuit (Chapter 4, *Electricity From a spinning shaft*) and makes for poor use of magnetic material. This is true, but the benefit is a simple design that's easy to construct—and at the end of the day we still wind up with an effective alternator that is a good match for the turbine blades. We could have used two smaller magnet rotors and stator, and a bit less magnetic material, and had the same results. But this way, most of the metal parts are interchangeable with our 10-foot turbine.

Just like with the 10-foot turbine, we use an aluminum template (Figure 20.6) and place 12 magnets around the steel rotor with opposite poles facing up. The magnets used are 2 inches in diameter by 1/2 inch thick, N40 or N42 grade NdFeB.

The single magnet rotor (Figure 20.7) mounts behind the wheel hub, just like with the 10-foot turbine. In front of that we mount a 6 inch diameter steel blade hub, with a 2-3/4 inch diameter hole in the center. This is the same part we use behind the blades on the 10-foot machine.

STATOR

The stator for the 7-foot turbine is identical to the one for the 10-foot turbine, except that we wind the coils with a different gauge of wire and different numbers of turns:

• 12 volts: 48 turns per coil using 2 strands of #15 AWG magnet wire.

• 24 volts: 90 turns per coil using a single strand of #15 AWG magnet wire.

• 48 volts: 190 turns per coil using a single strand of #18 AWG magnet wire.

Once the stator is finished, mount it to the stator bracket (Figure 20.8) and adjust for about 1/8 inch clearance between the stator and the magnet rotor. This should get you a cut-in speed of about 190 rpm. If you have a tachometer (see the sidebar), check the rpm at cut-in. If the cut-in speed is too low it will stall the blades, and that is easily fixed by adjusting the stator forward. The cut-in speed should be between 180 and 200 rpm.

Figure 20.8—Stator mounted in front of the single magnet rotor. There is no front magnet rotor on this 7-foot design.

TAIL

The tail is constructed just like the 10-foot wind turbine from all the same materials, but the tail boom is 4 feet long for this smaller turbine, and the tail bracket is 30 inches long. The shape of the wooden tail vane itself doesn't matter, but ours is a bit arrow shaped. It is 36 inches tall, 20 inches wide at the center and 16 inches wide across the top, and is approximately 4 square feet in area. See Chapter 5, *Furling and regulation*, for more about scaling the tail up or down.

BLADES

The blades are basically scaled down from the 10-foot design. We designed ours so they could be built from dimensional 2 x 6 lumber. They are 2 inches wide at the tips and full board width near the root, and the taper is straight (just like the 10-foot blades). The pitch at the tip is 3 degrees, at half the radius it's 6 degrees, and at the root we pitch the blades to full board

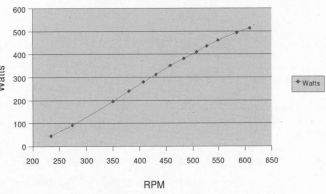

Rich's 7' wind turbine

Figure 20.9—Power versus rpm curve of the 7-foot turbine.

thickness. The airfoil at its thickest point is 1/8 as thick as the blade is wide. A bit thicker than that is OK, but you don't want to get thinner. This is a basic formula for a blade that works well. If it seems confusing, go back and read the instructions for the 10 foot diameter blades in Chapter 15, *Blades*.

CONCLUSION

We've only built three of these 7-foot turbines so far, but we are confident that it makes for a nice little 300 watt wind turbine. It's a bit less work and cost to build than the 10-foot turbine and the tower doesn't need to be as strong. Again, it is a bit goofy to run a single magnet rotor, but in practice it seems to work out fine. We also expect that the stator cools very nicely due to the fact that the wind can blow directly against it. The power versus rpm curve for the 7-foot machine is shown in Figure 20.9.

17-foot diameter, 3,000 watt wind turbine

At the time of writing we've built several of these (Figure 20.10), and the ones that have been installed have been holding up nicely. We often see sustained output of 3,000 watts or more, and peak output of over twice that. A machine this size on a good site can produce significant power and possibly run an entire energy-efficient home. It is however a much more expensive and difficult project than the 10 foot turbine. It is much too large for a 12 volt system and it would be fairly hard to handle at 24 volts. We only provide instructions on how to wind the stator for 48 volts, and recommend that this design be considered only for 48 volt systems.

Figure 20.10—Our 17-foot diameter wind turbine, scaled up from the 10-foot model.

METAL WORK

The metal work for the 17-foot turbine is very similar to the 10 foot machine. All of the critical angles are the same, but everything is both scaled up in size and strength. See Figure 20.11 for an index of the parts. Details about each part number are provided below. Again, these are not to be taken as plans, but a general description of what we have done. If you decide to experiment with building a larger machine, hopefully this will serve as somewhat of a guide. You may be able to make changes and possibly improvements based on your resources and intuition.

1) Stator bracket gussets. These are triangular pieces to stiffen the stator bracket. They're made from 1/4 inch steel, are 6 inches long, 1-1/2 inches tall at one end, and 1/2 inch tall at the other.

2) Wheel hub. This is a wheel hub for a trailer, and it's designed for a 6,000 pound axle. We get these from South West Wheel Company (see Chapter 21, *Sources*). It comes with bearings and uses a #42 spindle, which is available from the same company. This hub fits between the magnet rotors. The back rotor fits right up against the back of the hub, and often-

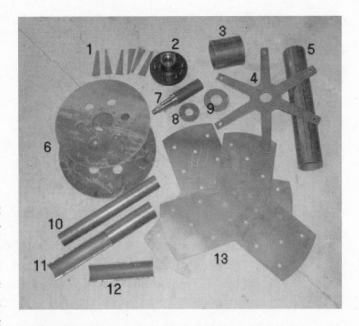

Figure 20.11—Metal parts for the 17-foot wind turbine.

times we have to turn the back of the hub slightly on a lathe so that the rotor will come flush to the back of the wheel hub.

3) Spindle housing. This is the bit that the stator bracket is welded to—it supports the spindle in its center. It's 5 inch diameter schedule 40 pipe, 5-3/4 inches long.

4) Stator bracket. It has 6 "arms" which support the stator, and is cut from 1/4 inch thick steel. The hole in the center is 2-1/4 inches in diameter. The holes at the end are 5/8 inch in diameter to allow for the 5/8 inch all thread studs we use to hold the stator on with.

5) Yaw bearing. The yaw bearing is the pipe that fits over the tower stub. It's 4 inch diameter pipe, 30 inches long.

6) Magnet rotors. These are cut from 1/2 inch thick steel and they're 18 inches in diameter. We put 6 holes (for cooling, and front rotor adjustment) around the center. They're 1-3/4 inches in diameter, evenly spaced around a 4 inch radius. Of course we also have 6 holes for the studs (5/8 inch in diameter) and 6 holes in the front rotor to be tapped 1/2-13 tpi for jacking screws. The hole in the center of the magnet rotors is 4-3/8 inch in diameter to accommodate the wheel hub.

7) Spindle. It's a standard #42 spindle that fits the hub—we also get that from South West Wheel company (see Chapter 21, *Sources*).

8) Yaw bearing cap. This is cut from 1/4 inch steel and fits in the top of the yaw bearing. It's 4 inches outer diameter and 1-1/4 inches inner diameter. The 4 inch diameter fits inside the top of the yaw bearing, and the 1-1/4 inch hole in the center is for the wires from the alternator to pass through into the tower.

9) Rear spindle support. This is the part that fits in the back of the spindle housing (part #3). It's 6 inches outer diameter (to fit inside the 6 inch pipe that the spindle housing is made from) and has a hole 2-1/4 inches diameter (so the spindle fits into it).

10) Tail pivot. It's 24 inches long made from 2 inch schedule 40 pipe.

11) Tail bearing. This is the part to which the tail boom is welded, and it rotates on the tail pivot as the machine furls. It has a notch cut out that is 9 inches long (from the bottom) and about half of the pipe (180 degrees) of the pipe is cut out—a bit less is OK or even better, it's not critical.

12) Bearing brace. This is the scrap from cutting out the notch for the tail bearing. We'll weld this back to the tail bearing to strengthen it.

13) Blade hubs. Overall they are 28 inches in diameter with 3 spars, 10 inches wide. We won't go into detail here. Obviously they need to have the 6 holes for the studs on a 5-1/2 inch circle to accommodate the studs. Then we have 5 more holes (5/8 inch diameter) for bolts to go through the blades. All these holes are 5/8 inch diameter. The rear blade hub has a 3 inch diameter hole in its center so that the grease cap on the hub can fit into it.

Missing from the picture above:

• **Tail boom.** Made of 1-1/2 inch pipe and 8 feet, 6 inches long.

• **Tail vane bracket.** This helps support the tail vane, and is 1-1/2 inch x 1/4 inch bar stock, 4 feet long.

• **Tail boom gusset.** This goes beneath the tail boom (it's welded on one end to the tail bearing and on the other end to the tail boom) and it's usually about 70 inches long.

• **Other gussets.** There are a couple of smaller gussets between the tail boom gusset and the tail boom to make things more rigid. We call those "tail boom gusset gussets."

• **Tail bracket.** This is the part that fits between the yaw bearing and the tail pivot, and is another triangular part cut from 1/2 inch thick steel. It's 9 inches long, 1 inch wide at the bottom and cut at 20 degrees. Figure 20.12 will hopefully clarify that a bit for you.

Figure 20.13 shows the whole alternator head pretty much finished. It also shows how the stator bracket gussets, stator bracket, spindle housing and rear spindle support all fit together. We cut a slightly oversized hole (about 4-1/2 inch diameter) into the side of the spindle housing to accommodate the short piece of 4 inch pipe that will attach the alternator to the yaw bearing. Again, it is all very much just like the 10 foot diameter turbine, but scaled up in size and much stronger.

Figure 20.12— Tail bracket being set up for welding to the tail pivot. To get things nicely centered, we mill a 3/8 inch slot into a piece of angle iron so that we're sure to get it perfectly centered and aligned with the tail pivot pipe.

Figure 20.14 shows the alternator head jigged up to weld to the yaw bearing. The above mentioned 4 inch pipe is overall 6 inches long, so that the alternator spindle is offset from the center of the yaw bearing (and the tower) by 8-1/2 inches. Look again at the 10-foot wind turbine plans (Chapter 9, *Frame*) to see how we do this without a jig. In the picture above we are using a jig we made—it makes life easier if you're making more than one machine but it's not necessary. Again, all of the angles are the same as the 10-foot turbine. The alternator head is tipped back 5 degrees to provide extra clearance between the blade tips and the tower. The geometry of the tail pivot with regard to the alternator and the yaw bearing is the same as on the 10-foot machine, too.

Figure 20.13—Finished alternator head.

Figure 20.14—Alternator head jigged up for welding to the yaw bearing.

Figure 20.15—Completed frame and tail boom.

TAIL

If you look over the plans for the 10 foot wind turbine again, the tail construction should be fairly obvious (Chapter 10, *Tail*). Again, this is all similar but just larger. The tail boom is 8 feet 6 inches long. The tail vane bracket is also scaled up. It's built from 2 inch wide x 1/4 inch thick steel bar stock, 50 inches long. The completed metal work is shown in Figure 20.15. Once the metal work is finished we either paint the machine, or (preferably) have it powder coated.

MAGNET ROTORS

The process for building the magnet rotors is the same as with the 10-foot machine, again—but bigger, heavier, and with much larger and more dangerous magnets! We have the rotors cut out first on a CNC CAD cutting machine. Then we drill holes for the studs, and tap the 6 holes for the jacking screws. We only need 3 holes for jacking screws, but we drill and tap 6 so that should one become stripped or damaged we have a redundant set. We usually then either paint or powder coat the magnet rotors before we place the magnets.

Using a template, we put the magnets down in the same way as the 10-foot wind turbine (Figure 20.16). It's also important to make an alignment mark on each rotor (just like the 10-foot wind turbine) so that you can be sure that both magnet rotors go together correctly with opposite poles directly opposing one another. These magnets are much larger, much more powerful and difficult to handle! Each rotor needs 16 N42 grade NdFeB magnets, 1-1/2 inch x 3 inch x 3/4 inch thick. This alternator contains over four times the magnetic material as the 10-foot diameter turbine. At the time of writing this, the cost of magnets to build the 17-foot machine is approximately $500.

We cannot stress enough how dangerous these magnet rotors are! It's important to

Figure 20.16—Magnets in place on the magnet rotor.

keep the rotors away from one another in a safe place where nobody will be surprised by them. They will do amazing things when you least expect it—like yank pocket knives out of pockets, grab tools from up to 18 inches away, and worse! Even a small metal tool flying onto a magnet rotor can cause injury if a finger gets in the way.

Figure 20.17—Banded magnet rotor. The steel rotor is 1/2 inch thick, and the magnets are 3/4 inch thick. The band allows 1/4 inch of the magnets to stick out, which allows for some airflow and improved cooling of the stator.

As described in Chapter 11, *Magnet rotors*, there is more than one way to cast and secure the magnets in place. We prefer to band the rotor edges (Figure 20.17) with stainless steel strap. With these rotors we start with 3/4 inch wide stainless steel banding material, and cut it 3/16 inch shorter than the circumference of the rotor. As described in that chapter, we heat the band and allow it to shrink around the rotor. The band is not as wide as the magnet rotor is thick, so the magnets protrude above the band by 1/4 inch. When running, this creates some airflow and improves cooling of the stator.

To prepare for pouring resin to make the castings, cut a disk 11 inches in diameter out of 3/4 inch plywood and run a bead of caulk around the edge. Then stick it to the center of the magnet rotor as shown in Figure 20.18. This temporary "island" serves as a dam to keep the resin where we need it when we cast the rotors.

Figure 20.18—Magnet rotor ready for casting.

Epoxy or vinyl ester are good resin choices for the magnet rotors. Mix it at about 50/50 with filler (talc or alumina tri-hydrate (ATH). It's also a good idea to mix in a bit of chopped fiberglass to add strength and to assure the casting will stay together if it ever cracks. Place the rotor on a level surface and carefully pour resin around the magnets, up to the height of the stainless steel strap (Figure 20.19). Once the resin has hardened, knock the plywood island out using a hammer through the hole in the center of rotor.

Figure 20.19—Pouring the resin.

Figure 20.20—Removing a coil from the larger coil winder.

Figure 20.21—Finished stator casting.

STATOR

The first step in building the stator is to build the coil winder. It's much like the one for the 10 foot wind turbine, but larger in all dimensions (Figure 20.20). The spool should be 6 inches in diameter. The coil winder should be constructed such that the coil will be 1/2 inch thick. The hole in the center needs to be 1 inch wide at the bottom, 1-1/2 inches wide at the top, and 3 inches tall. For a 48 volt system, we wind the coils with two strands of #14 AWG wire and 59 turns per coil. For 3-phase output we always use 3 coils per every 4 poles. Since this machine has 16 magnets (poles) per rotor, the stator requires 12 coils.

Wiring the stator is just like the 10 foot wind turbine. Every third coil is wired in series (so there are 4 coils per phase) and then the three phases are wired in star. The mold for casting the stator is 5/8 inch thick. The outer diameter is 22 inches, and the diameter of the island in the center is 9 inches. This is a much larger casting than the stator for the 10 foot machine, and it takes almost a gallon of resin mixture! With larger castings like this, it's wise to use less catalyst than the instructions call for. Large castings tend to get hot while curing, and if it gets too hot the casting can crack in the mold. So be sure to use plenty (50 percent by volume) of filler (talc or ATH) and go light on the catalyst. Figure 20.21 shows the finished stator casting.

ALTERNATOR ASSEMBLY

The 6 studs that hold the alternator together are cut from 5/8 – 11 tpi stainless steel threaded rod. Each stud is 7-1/2 inches long. Just like the 10-foot wind turbine, the back magnet rotor sits directly against the back of the wheel hub (see Figure 20.22). Use lock washers and jam nuts against the rotor. Jam nuts are not as tall, and are required behind the rotor due to lack of clearance between the rotor and the stator bracket. Be sure to get the hardware tight! We have also had clearance problems with some wheel hubs, depending on the hub—there may not be quite enough spindle protruding out of the alternator head. When that's the case we turn the wheel hub on a metal lathe. If you run into this problem and don't have a metal lathe, the back of the wheel hub can be taken down with a metal saw or a grinder.

Figure 20.22—Back magnet rotor mounted to the hub.

Grease the bearings and fit the back rotor and hub assembly to the alternator head. Adjust the bearings and give it a spin! Make sure the back magnet rotor doesn't wobble. If it runs true, install the grease cap.

The stator is supported by stainless steel 5/8-11tpi threaded rod. The procedure and hardware used for mounting the stator is the same as that on the 10-foot turbine (Chapter 14, *Alternator assembly*), just bigger, and there are 6 studs instead of 3. The studs that support the stator should be 5 inches long. Once the stator is in place, tighten the nuts on both sides of the stator. Leave the nuts at the stator bracket loose so that the stator can be centered between the magnet rotors later, after the front rotor is installed. The stator is shown mounted in Figure 20.23.

Figure 20.23—Stator mounted to the frame.

The front rotor will be supported against 6 nuts, one on each of the studs that protrudes from the wheel hub. Run one nut down on each stud all the way to the wheel hub. Be sure these nuts turn freely! Once the front magnet rotor is installed you'll have to adjust these with your fingers through larger holes in the front magnet rotor. If they don't turn freely, clean the threads and put a bit of oil on the nuts.

The front magnet rotor should be lowered carefully with 3 jacking screws. The jacking screws should be 10 inches long, cut from 1/2-13 tpi threaded rod. Put a drop of oil on each jacking screw. Run the jacking screws into the front magnet rotor so that 6 inches are protruding through the inside. It's wise to measure this and be sure that each jacking screw is protruding exactly the same amount, so that the rotor sits flat on them as soon as it's lowered into place.

Figure 20.24—Lowering the (very dangerous) front magnet rotor.

Make sure that the index marks on each magnet rotor are aligned so that you are positive the rotors will be in attraction with the poles directly opposing one another, and fit the front magnet rotor to the alternator. Carefully lower the front rotor (Figure 20.24) by turning each jacking screw not more than one turn at a time. Try to be sure it comes down flat, or else it will jam up against the studs. Sometimes a rubber mallet is useful to make sure the front rotor is always sitting against all 3 jacking screws. When the rotor gets close to the stator, the magnetic attraction becomes intense and you'll likely have to turn each jacking screw only 1/2 turn at a time.

With a machine this big, lowering the front rotor can be tricky and takes a bit of time and patience. Lower the front magnet rotor so that the air gap (the distance from the surface of magnets on the back rotor to the magnets on the front rotor) is exactly 1 inch. Turn the alternator and make sure the front rotor does not wobble—if it does, make adjustments with the jacking screws. We generally use a dial indicator to make sure the front rotor runs true.

ALTERNATOR TESTING

Once the air gap is 1 inch and you have checked that the front rotor runs true, check the cut-in speed. If you are spending all this money to build a 17-foot machine, you can afford a laser tachometer! Connect the alternator to a 3-phase rectifier to see at what rpm the output reaches 50 volts DC. It should be between 80 and 90 rpm. If it's more than 90 rpm, lower the front rotor slightly (decrease the air gap). If the cut-in speed is correct and the front rotor runs true, then reach through the holes in the front magnet rotor and run the extra nuts inside up to meet the front rotor with your fingers. Get them as tight as you can against the front rotor (you'll not likely get them very tight with your fingers, but that's OK). Then remove the 3 jacking screws. The front rotor should remain adjusted with the correct air gap, and come to rest tightly against the 6 nuts inside. Then run another nut down on each stud, tight against the front magnet rotor. The blade hubs will sit against these. The alternator is finished! The power curve is shown in Figure 20.27, next page.

Blades

The blades follow the same rules as other blades described in the book. Each blade is 8 feet 6 inches long, 6 inches wide at the tip, and 13 inches wide at the widest part of the root. We build them from laminated cedar, and overall the laminate is 2 inches thick (so that the thickest part of the blade that is supported by the hub is 2 inches). Like the other blades, the pitch at the tips is 3 degrees, at half the radius (4 feet 3 inches) it's 6 degrees, and at the root we pitch them to full board thickness. Unlike the smaller machines, we use a steel hub (described above) on each side of the blade set to hold them together. The hub has 6 holes in the center to fit the alternator, and then each blade gets 5 5/8-11 tpi bolts through it. Sometimes, before assembling the blades to the hub we'll use some construction adhesive on one hub for extra insurance that the blades stay in place.

The rest of the assembly and balancing should seem straight forward, it is just like the smaller machines (Figure 20.26).

Laser tachometers

These are one of the handiest tools we've ever seen for building and testing wind turbine alternators (Figure 20.9)! Stick a piece of reflective tape or a small, shiny magnet onto a spinning object. Hold the tachometer up close, hit the laser button, and read the rpm. They cost under $50, see the *Sources* chapter for electronics suppliers that carry them.

Figure 20.25—A laser tachometer.
Photo courtesy of All Electronics Corp.

Figure 20.26—Assembling the blades into a 17-foot rotor.

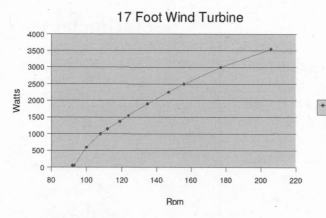

CONCLUSION

Again, this section should not be considered "plans," but hopefully there is enough detail so that the adventurer who chooses to try a larger system can build a workable machine. So far the 17-foot wind turbines we have built have done well. It is not nearly as easy to build as a 10-foot turbine, and the cost is much more. At the time of this writing it costs about $2,500 to build a 17-foot wind turbine, plus a good bit of your time!

Plan on spending at least this much on the tower, too. A tower to support a machine this size needs to be extremely robust. Even 6 inch diameter schedule 40 steel pipe would be marginal for the tower. We usually use 10 inch diameter 11 gauge tubing, which seems quite strong and has worked well for us. The 17-foot machine will easily produce 3 kw with sustained peaks over 6 kw, so you will need to size your conductors and rectifiers appropriately—prepare for peaks of over 125 amps at 48 volts.

WHY DON'T YOU...

We get questions like this all the time!

"Why don't you just put multiple stators and magnet rotors from the 10-foot turbine on the shaft of the 17-foot machine, instead of building this larger alternator?"

It's been done and lots of folks think about it. But it's not a good use of your magnet dollar. If you double the diameter of an alternator, and use twice the magnets and twice the copper, you will get 4 times the power at any given speed. This is because each coil is producing twice as much power when twice as many magnets go by per revolution and you have twice as many coils. If (for example) you double up with 2 stators, using twice the magnets and 3 magnet rotors (the middle one having magnets on both sides), you're using the same resources as you would have if you just doubled the diameter of the alternator—but you only get twice the power at any given rpm instead of 4 times the power.

The only good reason to do this is if there is some physical size restriction on the diameter of the alternator for your design.

---

21. Sources

It's very important to do as much homework as you can before even starting any sort of wind turbine project. The books and magazines listed below are excellent resources for anyone involved with wind power. And the quote at right says it all, too! There's no reason to re-invent the wheel when it comes to wind turbines.

Further reading:

"Wind Power Workshop" by Hugh Piggott. Centre for Alternative Technology Publications, 1997. A must-have book that covers every aspect of designing and building wind turbines at home.

"Axial Flux Alternator Wind Turbine Plans" by Hugh Piggott. Similar to our machines, just a different way to build them.

"Power With Nature, Second Edition" by Rex and LaVonne Ewing. Pixyjack Press, 2006. Everything you need to know about renewable energy power systems, but in a fun-to-read and friendly style.

"Got Sun, Go Solar" by Rex Ewing and Doug Pratt. Pixyjack Press, 2005. All about how to grid-tie renewable energy systems.

All of these books are available from us:
www.otherpower.com
Forcefield
2606 W Vine Dr
Fort Collins, CO 80521
(970) 484-7257 or (toll free in US) (877) 944-6247

"Wind Power - Renewable Energy for Home, Farm and Business" by Paul Gipe. The bible of wind power, from small systems to utility-scale machines.
Chelsea Green Publishing Co.
P.O. Box 428
White River Junction, VT 05001
(802) 295-6300

"Homeowner's Guide to Renewable Energy" by Dan Chiras, New Society Publishers, 2006. The basics of renewable energy systems for the homeowner who doesn't yet understand them.

"Power From the Wind" by Dan Chiras, with with Ian Woofenden and Mick Sagrillo. New Society Publishers, 2009. All about small-scale wind power system design and installation.

NOTABLE QUOTES

"A month in the laboratory can save you an hour at the library."
Unknown Fieldlines.com discussion board member

Both of these books available from:
New Society Publishers
P.O. Box 189
Gabriola Island, B.C.
Canada, V0R 1X0
1-800-567-6772 ext 111
orders@newsociety.com

Home Power Magazine. The best home-scale renewable energy magazine out there. Covers all aspects of solar, wind and hydro power from the perspective of the homeowner.
www.homepower.com
PO Box 520
Ashland, OR 97520
541-512-0201 or (toll free) 800-707-6585

Back Home Magazine. Renewable energy, plus homesteading, raising animals, and gardening too. They published the plans for our original Volvo-parts wind turbine design a few years back. It's still available as a PDF from them. Excellent magazine.
www.BackHomeMagazine.com
P.O. Box 70
Hendersonville, NC 28793
800-992-2546

Backwoods Home Magazine. Homesteading from a self-reliance and survivalist perspective. Another great magazine.
www.backwoodshome.com
P.O. Box 712
Gold Beach, OR 97444
1-800-835-2418

The Mother Earth News. Green living, gardening and renewable energy since 1970!
www.motherearthnews.com
1503 SW 42nd St.
Topeka, Kansas 66609-1265
1-800-234-3368

Homebrew wind turbine parts and kits:

Figure 21.1—Cad-cut flat metal parts kit.

Figure 21.2—Aluminum magnet rotor layout template.

We can provide you with precise CAD-cut metal parts for building the 10 foot wind turbine in this book. The circle cuts are perfect, all the drill holes are aligned to fit the trailer hub used in this book and are tapped, and we can even powder coat certain parts for you. Examples of what's available are shown here, but be sure to check our website for an up to date list. Buying prefabricated parts will save you time, but cost you more money.

• Complete wind turbine kit. Blade set, frame, and fully assembled alternator. All you do is paint it, mount the rotor, and fly it.
• Finished metalwork kit. All the metal parts, with the frame already welded together for you.
• Flat metal parts kit (Figure 21.1).
• Magnet rotor template (Figure 21.2).
• Pre-wired and cast stators.
• Blade set.
• Magnet wire.
• Magnets.
• Rectifiers.
• Stainless steel hardware.

www.otherpower.com
Forcefield
2606 W Vine Dr
Fort Collins, CO 80521
(970) 484-7257 or (toll free in US) (877) 944-6247

Anemometers

Professional grade logging anemometers that use SD memory cards for data storage, free open-source logging software, and a really cool wind turbine performance data acquisition center.
APRS World LLC
www.aprsworld.com/
PO Box 1264
Winona MN 55987
507-454-2727

Professional grade logging anemometers and accessories.
NRG Systems, Inc.
www.nrgsystems.com
PO Box 0509
Hinesburg, Vermont 05461
802-482-2255

Inexpensive anemometers that use bicycle computers to log wind speeds and wind miles:
Clean Energy Products
cleanenergy@hotmail.com
Box 9413
Bend, OR 97708
206-953-4039

Inspeed
www.inspeed.com
INSPEED.COM, LLC
10 Hudson Road
Sudbury, MA 01776
(978) 397-6813

Our webpage about building a bicycle speedometer anemometer. We also sell the cup and hub set for it:
www.otherpower.com/anemom2.html

Diversion load controllers:

Xantrex C-series controllers work very well with our wind turbine design. Widely available. To find a distributor or retailer near you, contact:
Xantrex Technology Inc.
8999 Nelson Way
Burnaby, BC Canada V5A 4B5
604-422-8595

Morningstar also makes an excellent controller that works very well with our wind turbines. Also widely available. To find a distributor or retailer near you, contact:
Morningstar Corporation
www.morningstarcorp.com
1098 Washington Crossing Road
Washington Crossing, PA 18977
215-321-4457

Dump load heating elements:

Northwest Power Co, LLC
www.nwpwr.com
P.O. Box 271
Platteville, CO 80651
970-785-2707

In addition to supplying dump load heating elements, both of these companies below are also full-line dealers in solar panels, controllers, batteries, and everything else needed for a renewable energy system:

The Alternative Energy Store
www.AltEnergyStore.com
43 Broad Street, Suite A408
Hudson, MA 01749
(877) 878 - 4060 or (978) 562-5858
Costa Rican office: +506 297 14 04.

Backwoods Solar Electric Systems
www.backwoodssolar.com
1589 Rapid Lightning Creek Rd
Sandpoint, ID 83864
208-263-4290

Electronic components:

(rectifiers / switches / ammeters / heat sinks / plugs and sockets / resistors)

All Electronics Corp.
www.allelectronics.com
14928 Oxnard St.Van Nuys, CA 91411-2610
818-904-0524 or 888-826-5432

Digikey
www.digikey.com
701 Brooks Avenue South
Thief River Falls, MN 56701
800-344-4539 or 218-681-6674

MECI Liquidation Outlet
www.meci.com
340 E. First St.
Dayton, OH 45402
800-344-4465

A great directory of surplus electronics suppliers:
www.amasci.com/supliers.html

Hubs, bearings and spindles:

Etrailer.com
www.etrailer.com
1507 East Hwy A
Wentzville, MO 63385
1-800-298 8924

SouthWest Wheel Company
www.southwestwheel.com
1501 E Broadway
Lubbock, TX 79403
1-800-866-3339

Casting resins, fiberglass cloth, and filler:

US Composites
www.uscomposites.com
5101 Georgia Avenue
West Palm Beach, FL 33405
561-588-1001

Wind turbine tower kits:

Bergey WindPower Co.
www.bergey.com
2200 Industrial Blvd.
Norman, Ok 73069
(405) 364-4212

Southwest Windpower
www.windenergy.com
1801 W. Route 66
Flagstaff, AZ 86001 USA
928-779-9463
IDC Solar
www.idcsolar.com
PO Box 630
Chino Valley, AZ 86323 USA
(928) 636-9864

We highly recommend talking to the folks in Sturgeon Bay below about tower kits! They can also fabricate tower-top stubs for you:

Lake Michigan Wind and Sun
www.windandsun.com
1015 County Rd U
Sturgeon Bay WI 54235-8353
920.743.0456

Independent Power Systems (IPS)
www.solarips.com
1501 Lee Hill Rd #19
Boulder, CO 80304
303-443-0115

The Energy Depot Inc.
www.energydepot.ca
16650 Jane St.
Kettleby ON L0G-1J0 Canada
905-760-7511

Abundant Renewable Energy LLC
www.abundantre.com/towers.htm
22700 NE Mountain Top Road
Newberg, Oregon 97132
(503) 538-8298 or (503) 883-1003 (Sales)

Earth anchors:

To see some typical guy wire anchors, go to:
www.hubbellpowersystems.com/POWERTEST/chance/earth_anchors.html
To find a distributor near you, go to:
www.abchance.com/ch_dist.html

Rigging:

(wire rope, cable clamps, turnbuckles, thimbles, etc.)
Web Rigging Supply, Inc
www.WebRiggingSupply.com
27W966 Commercial Ave.
Lake Barrington, IL 60010
(877) 744-4461 or (847) 304-4550

Lattice towers:

www.criticaltowers.com/
Sandown Wireless
P.O.Box 564
East Hampstead, NH 03826
866-379-8437 or 603-974-0725

Commercial wind turbines:

It's very interesting, entertaining, and informative to read about the huge variety of small commercial wind turbines that are available for purchase. And many of their websites have detailed information about the towers needed to safely fly their turbines. Note that turbines listed as "off grid" can still be grid tied through a battery bank and grid tie inverter. "Direct grid tied" turbines generally cannot be used with a battery bank.

Abundant Renewable Energy
2.5 kw and 10 kw turbines, both off grid and direct grid tied.
www.abundantre.com
22700 NE Mountain Top Road
Newberg, Oregon 97132, USA
503 538-8298 (Telephone)
503 538-8782 (Fax)

Aerostar, Inc.
10 kw direct grid tied turbine.
www.aerostarwind.com
PO Box 52
Westport Point, MA 02791
info@aerostarwind.com
508-636-5200

Bergey WindPower Co.
1 kw, 1.5 kw and 10 kw turbines, both off grid and direct grid tied.
www.bergey.com
2200 Industrial Blvd.
Norman, Ok 73069 USA
(405) 364-4212

Bornay Wind Turbines
600 w, 1.5 kw, 3 kw, and 6 kw turbines, off grid and direct grid tied.
www.bornay.com
P.I. RIU, Cno. del Campanar
03420 Castalla (Alicante) Spain
Tel. 965 560 025 – 966 543 077

Chinook Wind Turbines
250 w, battery charging.
www.chinookturbines.com
info@chinookturbines.com

Flowtrack Australia
5 kw turbine, both off grid and direct grid tied.
www.flowtrack.com.au
FLOWTRACK Pty Ltd

0266 891431 (a/h)
0266 890408 (b/h)
kali@nimnet.asn.au
End of Tuntable Falls Road, NIMBIN
Australia

Iskra Wind Turbines Ltd.
5 kw turbine, both off grid and direct grid tied.
www.iskrawind.com
enquiries@iskrawind.com
The Innovation Centre
Epinal Way
Loughborough LE11 3EH
United Kingdom
0845 8380588

Jacobs Wind Turbines
Wind Turbine Industries Corp.
10 kw, 12.5 kw, 15 kw and 17.5 kw direct grid tied.
www.windturbine.net
16801 Industrial Circle S.E.
Prior Lake, Minnesota 55372 USA
952-447-6064

Kestrel Wind Turbines
600 w, 800 w, 1 kw, and 3 kw, both off-grid and
direct grid tied.
www.kestrelwind.co.za
Eveready, Eveready Road
Struandale, Port Elizabeth South Africa.
+27 (0)41 401 2645 / 2500
kestrelwind@eveready.co.za

Marlec Engineering Co Ltd.
60 w, 144 w, 250 w, and 720 w, off grid.
www.marlec.co.uk
Rutland House -Trevithick Rd - Corby
Northants -NN17 5XY - United Kingdom
+44 (0)1536 201588

Proven Energy Ltd.
2.5 kw, 6 kw and 15 kw, both off grid and direct grid
tie.
www.provenenergy.co.uk
Wardhead Park
Stewarton, Ayrshire, KA3 5LH
Scotland, UK
0044 (0) 1560 485 570
info@provenenergy.com

Southwest Windpower
www.windenergy.com
400 w, 900 w, 1 kw, 1.8 kw, and 3 kw, both off grid
and direct grid tie.

1801 W. Route 66
Flagstaff, AZ 86001 USA
928-779-9463
info@windenergy.com

West Wind
3 kw, 5 kw, 10 kw, and 20 kw, both off grid and direct
grid tied.
www.westwindturbines.co.uk/
J.A Graham Group
3 Carmavy Road, Nutts Corner
Crumlin Co., Antrim BT29 4TY
Northern Ireland
+44 28 94 452 437
info@jagraham.com

Goverment agencies

National Renewable Energy Laboratory (NREL)
www.nrel.gov
1617 Cole Blvd.
Golden, CO 80401-3393
(303) 275-3000

Critical websites

Hugh Piggott's website
Hugh can truly be called the father of homebrewed
wind power. Every one of his books and sets of plans
should be considered essential reading for anyone
contemplating building a wind turbine. His website
contains a huge wealth of free information:
www.scoraigwind.com

Mike Klemen's website
Mike's detailed data acquisition over many years and
with many different turbines has provided the wind
power world with a huge amount of data on what
wind turbines can really do, and what they can't:
www.ndsu.nodak.edu/ndsu/klemen/

Paul Gipe's website
Paul is one of the most prolific and respected wind
power authors worldwide, and his monumental book
"Wind Power" is the tome we referred to over and
over to check our facts here. His website is chock full
of articles about small-scale wind, detailed
measurements and performance analysis of numerous
small wind turbines, and the real facts and data about
urban and rooftop wind power myths. Paul's entire
website, and his books, are essential reading.
www.wind-works.org/

American Wind Energy Association (AWEA)
Non-profit wind power advocacy group, covering the

entire wind power industry. Lots of free information. Their support of small wind power is impressive, and they have detailed and frequently updated information on the status of wind power legislation in the US, including net metering and grid tie laws. *http://www.awea.org/*

Internet discussion:

We highly recommend checking out our homebrew energy discussion board at:
www.fieldlines.com
Here you'll find 10,000+ people from all over the world interested in do-it-yourself renewable energy systems. Many users are actively building and flying their own wind turbines. Well worth checking out, and a great online community.

Contact us:

Forcefield
2606 W Vine Dr

Fort Collins, CO 80521
(970)-484-7257 or (toll free in US) 877-944-6247
www.otherpower.com - renewable energy
www.wondermagnet.com - magnets
www.forcefieldmagnets.com - online store for all of our products
www.homebrewwind.com - up-to-date information about this book, including critical updates, improved construction techniques, and random wind power rambling and rants by the authors.
windturbine@otherpower.com - a great way to get in touch us, so we'll know your inquiry is about wind power.

Please keep in mind that there is no phone service (or even cell phone reception) at our wind turbine shop up in the mountains! The phone numbers shown are for our shipping department in town, and there may or may not be someone available there who is able to answer your renewable energy questions. Email is by far the best way to contact us.

Hugh Piggott: homebrew wind power pioneer

If it were not for the patient and kind help we've received over the years from Hugh Piggott, the authors would still be floundering around with turbines that break easily and perform poorly, and this book would never have been written. Hugh was born in Scotland in 1952 and educated at Edinburgh and Cambridge, then moved straight to the remote Scoraig peninsula in NW Scotland to live a very basic lifestyle growing vegetables, cows and such. He is married with two children. Scoraig has good wind exposure and no utility power lines, so windpower was a natural choice to supply electricity for lighting. With help from lots of people, Hugh designed and built small wind turbines made from recycled parts that supplied local families with basic 12 volt lighting. In 1990 he started to teach courses at the Centre for alternative Technology (CAT) in Wales. His wind turbine ideas progressed over thirty years and Hugh designed larger turbines for third-world manufacture, wrote books and taught wind turbine construction courses in several countries. Much of this work is documented on the web pages at *www.scoraigwind.com*. Several groups around the world use Hugh's designs for their own products and courses. He has installed various renewable energy systems using wind, solar and hydro power but mainly works as a teacher, writer and consultant based at Scoraig. He spends a lot of time answering emails from wind turbine enthusiasts. Hugh's publications include:

- *"Scrapyard Windpower Realities"* mid-1980s by CAT (now only sold as a pirate edition)
- *"Windpower Workshop"* 1997 by CAT publications
- *"The Brakedrum Windmill Plans"* 2000 by Picoturbine (now out of print)
- *"Small Wind Systems for Rural Energy Services"* co-authored 2003 ITDG publishing
- *"How to Build a Wind Turbine"* 2005 self published
- *"Choosing Windpower"* 2006 by CAT publications
- *"A Wind Turbine Recipe Book"* 2008 self published

22. Glossary

Ø—Symbol for Diameter

AC—See Alternating Current

Airfoil—The cross section profile of a wind generator blade. Designed to give low drag and good lift. Also found on an airplane wing.

Air Gap—In a permanent magnet alternator, the distance between the magnet rotors.

Alternating Current—Electricity that changes direction periodically. The period is measured in Cycles per Second (Hertz, Hz).

Alternator—A device that produces Alternating Current from the rotation of a shaft.

Amperage—A unit of electrical current, equal to Coulombs per second. This is the flow rate of electrons moving through a circuit, very roughly analogous to gallons per minute flowing from a faucet.

Ampere-Hour—A measure of energy over time, equal to amperes times hours. Also used to measure battery capacity.

Anemometer—A device that measures wind speed.

Angle of Attack—The angle of relative air flow to the leading edge of the blade.

Apparent wind—The wind direction and speed that the leading edge of a wind turbine blade encounters. Since the blade tips are likely moving faster than the wind (see Tip Speed Ratio), apparent wind is both faster and from a different direction than the actual wind.

Area of a Circle—Pi multiplied by the Radius squared.

Armature—The moving part of an alternator, generator or motor. In many PM alternator designs, it carries the magnets and is attached to the blades and hub. In this book, we call it the "Magnet Rotor" instead.

Axial Flux Alternator—Alternator design where a flat disc with magnets on the face rotates around a flat disc carrying coils (the Stator), parallel with the shaft.

Axis—The centerline around which something rotates.

Balancing—With wind turbine blades, adjusting their weight and weight distribution through 2 axes so that all blades weight the same in all dimensions. Unbalanced blades create damaging vibration.

Battery—An electrochemical device for storing energy.

Battery Bank—An array of Batteries connected in series, parallel, or both.

Bearing—A device that transfers a rotating force to the Spindle.

Beer—magical fluid consumed by some wind turbine enthusiasts after the day's work is done.

Betz Coefficient—59.26 percent. This is the theoretical maximum efficiency at which a wind generator can operate, by slowing the wind down. At the Betz limit in power extraction, the wind would 'rather' go around the wind turbine instead of through it.

Bird—A small flying creature that is far more likely to be killed by a building or a cat then by a wind turbine.

Blade—The part of a wind generator rotor that catches the wind. It has flat faces towards the wind, and an airfoil on the back.

Braking System—A device to stop a wind turbine from spinning. In this design, it's an electrical brake.

Bridge Rectifier—An array of diodes used to convert Alternating Current to Direct Current. Single-phase bridge rectifiers use 4 diodes, 3-phase bridge rectifiers use 6 diodes.

Brushes—Devices for transferring power to or from a rotating object. Usually made of carbon-graphite.

Cable Clamp—A clamp that holds two pieces of guy wire together, 3 are used to make a loop for attachment.

Ceramic Magnets—See Ferrite Magnets.

Chord—The width of a wind turbine blade at a given location along the length.

Circuit Breaker—A switch that automatically shuts off if too much current is flowing in a circuit. See also Fuse.

Coercivity—The amount of power needed to magnetize or demagnetize a permanent magnet, in Mega-Gauss Oersted (mGOE).

Coffee—magical fluid that enables wind turbine enthusiasts to function early in the morning.

Cogging—The cyclic physical resistance felt in some alternator designs from magnets passing the coils and gaps in the laminates. Our turbine design here does not cog.

Coil—A length of wire wound around a form in multiple turns.

Commutator—The part in some generators and alternators that transmits electricity to and from moving parts. All DC generators have commutators, some alternators do. PM alternators like the one featured in this book do not.

Concave—A surface curved like the interior of a circle or sphere.

Contactor—A heavy-duty Relay.

Controller—A device that prevents a battery bank from overcharging. Wind turbines need a special kind, called a Diversion Controller.

Convex—A surface curved like the exterior of a circle or sphere.

Cowling—See Nacelle.

Current—See Amperage.

Cut-In—The RPM at which an alternator or generator starts pushing electricity hard enough (has a high enough voltage) to make electricity flow in a circuit. The alternators here are said to be at cut-in when their voltage exceeds the battery bank voltage.

Cyanoacrylate—A fast-setting, hard and brittle adhesive. See Superglue®.

Cycles per Second—Measured in Hertz. In electricity, it is the number of times an AC circuit reaches both minimum and maximum values in one second.

Darrieus—A Vertical Axis Wind Turbine design from the 1920s and 1930s by F.M. Darrieus, a French wind turbine designer.

DC—See Direct Current

Dead End—The short end of a guy wire.

Delta—One possible configuration for wiring 3-phase alternators and motors, a sort of parallel connection.

Diameter (Ø)—The length of a straight line passing through the center of a circle, and ending on both edges. Equal to 2 times the Radius.

Diode—A solid-state device that allows electricity to flow in only one direction.

Diversion Controller—A Controller that takes power directly from a battery bank and diverts it to heating elements to prevent batteries from overcharging.

Diversion Load—see Dump Load.

Dog—1) A 4-legged creature essential to the construction of Wind Turbines. 2) Your best friend.

Downwind—Refers to a Horizontal Axis Wind Turbine in which the hub and blades are downwind from the tower, the opposite of an Upwind turbine. Downwind turbines do not need Vanes, upwind turbines do.

Drag—In a wind generator, the force exerted on an object by moving air. Also refers to a type of wind generator or anemometer design that uses cups instead of a blades with airfoils. Good turbine designs minimize drag and maximize lift.

Dump Load—A device to which wind generator power flows when the system batteries are too full to accept more power, usually an electric heating element. This diversion is performed by a Diversion Controller, and allows a Load to be kept on the Alternator or Generator at all times.

Duty Cycle—In a circuit, the ratio of on time to off time.

Dynamo—A device that produces Direct Current from a rotating shaft. See Generator.

Eddy Currents—Unwanted currents that flow in a conductor from variations in magnetic induction. See also Lenz Effect.

Efficiency—The ratio of energy output to energy input in a device.

Electromagnet—A device made of wire coils that produces a magnetic field when electricity flows through the coils.

Epoxy—A 2-part adhesive system consisting of resin and hardener. It does not start to harden until the elements are mixed together.

Fatigue—Stress that causes material failure from repeated, cyclic vibration or force.

Fat Tire—see Beer. The kind we drink here.

Ferrite (Magnets)—Also called Ceramic Magnets. Made of Strontium Ferrite. High Coercivity and Curie Temperature, low cost, but brittle and 4-5 times weaker than NdFeB magnets.

Field—See Magnetic Field.

Flux (Magnetic)—See Magnetic Field.

Freewheeling—a wind generator that is not connected to a Load. It is in danger of self-destruction from overspeeding.

Frequency—See Cycles per Second.

Furling—The act of a wind generator Yawing out of the wind to protect itself from high wind speeds.

Furling Tail—A wind generator protection mechanism where the rotor shaft axis is offset horizontally from the yaw axis, and the tail boom is both offset horizontally and hinged diagonally, thus allowing the tail to fold up and in during high winds. This causes the blades to turn out of the wind at high wind speeds, protecting the machine.

Fuse—Like a Circuit Breaker, but must be replaced if it stops the current flow.

Gauss—A unit of magnetic field strength, equal to 1 Maxwell per square centimeter. Higher Gauss measurements mean more power can be induced to flow in an alternator.

Gearing—Using a mechanical system of gears or belts and pulleys to increase or decrease shaft speed. Power losses from friction are inherent in any gearing system.

Generator—A device that produces Direct Current from a rotating shaft.

Gin Pole—A lever arm that gives a mechanical advantage when erecting a tower.

Guy Anchor—Attaches tower guy wires securely to the earth.

Guy Radius—The distance between a wind turbine tower and the guy anchors.

Guy Wire—Attaches a tower to a Guy Anchor and the ground.

H-Rotor—A Vertical Axis Wind Turbine design.

HAWT—See Horizontal Axis Wind Turbine.

Hertz—Frequency measurement. See Cycles per Second.

Horizontal Axis Wind Turbine (HAWT)—A 'normal' wind turbine design, in which the shaft is parallel to the ground, and the blades are perpendicular to the ground.

Hub—The center of a wind generator rotor, which holds the blades in place and attaches them to the shaft.

Impedance—See Resistance.

Induction(Magnetic)—The production of electricity by the movement of a magnetic field past a wire.

Inverter—A device that converts DC electricity into AC.

Jin Pole—See Gin Pole.

Kerf—The width of a cut made by a saw.

Kilowatt—1000 Watts (see Watt)

Kilowatt-Hour—A unit of energy, equalling one Kilowatt of power made or used for one hour.

kw—See Kilowatt.

kwh—See Kilowatt-Hour.

Laminations—Electrical circuit core parts, found in motors, generators, alternators and transformers. Made of high silicon steel, insulated from each other, and designed to prevent Eddy Current losses.

Lattice—In wind power, a type of tower made from many struts.

Leading Edge—The edge of a blade that faces toward the direction of rotation.

Leeward—Away from the direction from which the wind blows.

Lenz Effect—See also Eddy Currents. From H.F.E Lenz in 1833. Electromotive force is induced with variations in magnetic flux. It can be demonstrated physically in many different ways—for example dragging a strong magnet over an aluminum or copper plate, or shorting the terminals of a PM alternator and rotating the shaft by hand. Steel laminates are used to reduce power losses from this effect.

Lift—The force exerted by moving air on asymmetrically-shaped wind generator blades at right angles to the Apparent Wind. Ideally, wind generator blades should produce high Lift and low Drag.

Live—A circuit that is carrying electricity. When live, it can shock you.

Load—Something physical or electrical that absorbs energy. A wind generator that is connected to a battery bank is loaded.

Losses—Power available in the wind but is not transferred to a usable form. Losses can be from friction, electrical resistance, the fluid dynamics of air, or other causes.

Magnet—A body that attracts ferromagnetic materials. Can be a Permanent magnet, Temporary Magnet, or Electromagnet.

Magnetite—A common Iron-containing mineral with ferromagnetic properties.

Magnet Wire—The kind of wire always used in making electromagnets, alternators, generators and motors. Uses very thin enamel insulation to minimize thickness and maximize resistance to heat.

Magnetic Circuit—The path in which magnetic flux flows from one magnet pole to the other.

Magnetic Field—Magnetic fields are historically described in terms of their effect on electric charges. A moving electric charge, such as an electron, will accelerate in the presence of a magnetic field, causing it to change velocity and its direction of travel. An electrically charged particle moving in a magnetic field will experience a force (known as the Lorentz force) pushing it in a direction perpendicular to the magnetic field and the direction of motion. Also called magnetic flux.

Maker's Mark—see Whiskey. The kind we use here after all work is done, and the day's wind turbine projects were stressful.

Maximum Energy Product—Determines how good a magnet that different materials can make. Technically, the amount of energy that a material can supply to an external magnetic circuit when operating within its demagnetization curve. Also called the 'Grade' of a magnetic material, and measured in megaGauss Oerstead (MGOe).

MegaGauss Oersted—Magnetic force measurement, see Maximum Energy Product.

MGOe—See MegaGauss Oersted.

Moment—A force attempting to produce motion around an axis.

Monopole—In wind power, a Freestanding, tubular tower with no guy wires. In magnetism, a fictional magnet that has only one pole. See also Poles.

Mouse—In wind power, the act of threading wire through the tower turnbuckles so they cannot loosen from vibration.

NdFeB—See Neodymium-Iron-Boron (Magnet).

Nacelle—The protective covering or cowling over a generator or motor.

Neodymium-Iron-Boron (Magnet)—The composition of the most powerful Permanent Magnets known to mankind. The materials are mined, processed, pressed into shape, and sintered. Then, they are subjected to an extremely strong magnetic field and become Permanent Magnets.

Ohm's Law—The basic math needed for nearly all electrical calculations.

Open-Circuit Voltage—The voltage that a alternator or generator produces when it is NOT connected to a Load.

Parallel—In DC electrical circuits such as a battery bank or solar panel array, this is a connection where all negative terminals are connected to each other, and all positive terminals are connected to each other. Voltage stays the same, but amperage is increased. Opposite of Series.

Permanent Magnet—A material that retains its magnetic properties after an external magnetic field is removed.

Permanent Magnet Alternator—An Alternator that uses moving permanent magnets instead of Electromagnets to make the magnetic field.

Personal Protective Equipment (PPE)—Articles of protective clothing that humans (and sometimes dogs) wear to protect themselves from workshop hazards. Includes eye, ear, lung, and skin protective equipment.

PM—See Permanent Magnet.

PMA—See Permanent Magnet Alternator.

Phase—The timing of AC current cycles in different wires. 3-phase alternators produce current that is cyclically timed between 3 different wires. In a 3-phase alternator, wire #1 receives a voltage peak, then wire #2 receives a peak, then wire #3, and so on.

Pitch—See Setting Angle.

Poles—A way of picturing magnetic phenomena. All magnets are considered to be 'dipoles,' having both a North pole (which would point North if used in a compass) and a South pole (which would point South if used in a compass).

PPE—Personal Protective Equipment. The stuff you wear to prevent yourself (hopefully) from being killed or maimed!

Prop—Slang term for Propeller.

Propeller—The spinning thing that makes an airplane move forward. Often incorrectly used (by the authors also!) to describe a wind turbine Rotor.

Radius—The distance between the center of a circle and the outside.

Rare-Earth Magnets—See Neodymium-Iron-Boron magnets.

Rated Power Output—Used by wind generator manufacturers to provide a baseline for measuring performance. Rated output may vary by manufacturer. For example, one manufacturer's 1500 watt turbine may produce that amount of power at a 20 mph windspeed, while another brand of 1500 watt turbine may not make 1500 Watts until it gets a 40 mph windspeed. So read manufacturer's ratings statements very carefully.

Rayleigh Distribution—A Weibull Distribution with the shape parameter=2.

Rectifier—See Diode.

Radial—Alternator design where the magnets rotate perpendicular to the shaft.

Regulator—See Controller.

Relay—An electromechanical switch that uses a small amount of incoming electricity to charge an electromagnet, which physically pulls down a connecting switch to complete a circuit. This allows a low-power circuit to divert the electricity in a high-power circuit.

Resin—A two-part compound that sets with an exothermic reaction to form a solid block of plastic.

Resistance—The resistance to current flow in an electrical circuit, measured in Ohms. See Ohm's Law.

Root—The area of a blade nearest to the hub. The thickest and widest part of the blade.

Rotor—1) The blade and hub assembly of a wind generator. 2) The Armature of a permanent magnet alternator, which spins and contains permanent magnets.

RPM—Revolutions Per Minute. The number of times a shaft completes a full revolution in one minute.

Savonius—A vertical-axis wind turbine (VAWT) design by S.J. Savonius of Finland from the 1920s and 30s. Shaped like a barrel split from end to end and offset along the cut. They are drag machines, and thus give very low rpm but lots of torque. Not generally used for making electricity.

Series—In DC electrical circuits such as a battery bank or solar panel array, this is a connection where all the negative terminals are connected to the neighboring positive terminals. Voltage increases, but amperage stays the same. Opposite of Parallel.

Servo Motor—A motor used for motion control in robots, hard disc drives, etc. Generally designed more like an alternator than a standard motor, most Servos need special control circuitry to make them rotate electrically. Some can be used in reverse to generate alternating current.

Setting Angle—The angle between the blade Chord and the plane of the blade's rotation. Also called Pitch or blade angle. A simple blade has the same setting angle along it's entire length. A blade carved with a Twist has a different setting angle at the Tip than at the Root.

Shackle—A U-shaped piece of metal secured with a bolt, for connecting guy wires to anchors and towers.

Shaft—The rotating part in the center of a wind generator or motor that transfers power.

Short Circuit—Parts of a circuit connected together with only the impedance of the leads between them. In wind generators, connecting the output leads directly together so as to heavily load a generator in extreme winds, thus shutting it down.

Shunt—An electrical bypass circuit that proportionally divides current flow between the shunt and the shunted equipment. It allows high current measurements with low-current equipment.

Shunt Regulator—A bypass device for power not needed for charging batteries. When batteries are full, the regulator shunts all or part of the excess power to a Dump Load to protect the batteries from overcharging damage.

Slip Ring—A device used to transfer electricity to or from rotating parts. Used in wound-field alternators, motors, and in some wind generator yaw assemblies.

Spindle—A stationary part on which a bearing rides, allowing an assembly to rotate.

Star—One possible way to connect the phases in a 3-phase Alternator, a kind of Series connection. Also called a Wye connection.

Start-Up—The windspeed at which a wind turbine rotor starts to rotate. It does not necessarily produce any power until it reaches Cut-In speed.

Stationary—With wind generator towers, a tower that does not tilt up and down. The tower must be climbed or accessed with a crane to install or service equipment at the top.

Stator—The part of a motor, generator or alternator that does not rotate. In permanent magnet alternators it holds the coils.

Superglue®—Trademark name for Cyanoacrylate adhesive. Fast bonding glue, easy to find in different viscosities. Sets on its own, and sets instantly when sprayed with an accelerator chemical. Hard, but somewhat brittle.

Tail—See Vane. The proper term is actually Vane, but Tail is commonly used. The part of a wind turbine that sticks out the back and keeps it aligned with the oncoming wind.

Tail Boom—A strut that holds the tail (Vane) to the wind generator frame.

Tape Drive Motor—A large permanent magnet DC motor from large, old surplus computer tape drives. Sometimes used as a generator for building tiny wind turbines for science fair projects and the like.

Taper—The change in wind turbine blade width (chord) along the length.

Temporary Magnet—A material that shows magnetic properties only while exposed to an external magnetic field.

Tesla—Equal to 10,000 Gauss.

Thimble—A metal channel inserted inside the loop of a guy wire, to prevent abrasion.

Thrust—In a wind generator, wind forces pushing back against the rotor. Wind generator bearings and towers must be designed to handle thrust or else they will fail.

Thrust Bearing—A bearing that is designed to handle axial forces along the centerline of the shaft—in a wind generator, this is the force of the wind pushing back against the blades.

Tilt-Up—A tower that is hinged at the base and tilted up into position using a gin pole and winch or vehicle. Wind turbines on tilt-up towers can be serviced on the ground, with no climbing required.

Tip—The skinny end of a wind generator blade farthest from the hub.

Tip Speed Ratio—The ratio of how much faster than the windspeed that the blade tips are moving. Abbreviation=TSR.

Torque—Turning force, equal to force times radius. See also Moment.

Tower—A structure that supports a wind generator, hopefully high in the air.

TPI—Threads per inch, on a bolt or screw.

Trailing Edge—The edge of a blade that faces away from the direction of rotation.

Transformer—Multiple individual coils of wire wound on a laminate core. Transfers power from one AC circuit to another using magnetic induction. Used to step voltage up or down.

TSR—See Tip Speed Ratio.

Turn—In winding stator coils, this is one loop of wire around a form. A coil will often be referred to by how many turns of a certain gauge wire are in each coil.

Turnbuckle—A device with two screw ends that allows you to adjust the length of a guy wire.

Twist—In a wind generator blade, the difference in Pitch between the blade root and the blade tip. Generally, the twist allows more Pitch at the blade root for easier Startup, and less Pitch at the tip for better high-speed performance. A properly-designed blade is twisted so that when running at the design Tip Speed Ratio, all parts of the blade are presenting the correct angle of attack to the wind.

Upwind—A wind turbine design where the blades are upwind of the tower. Small upwind turbines use a Vane to face into the wind, large utility-scale turbines use electronic controls and servo motors.

Vane—A large, flat piece of material used to align a wind turbine rotor correctly to face the wind. Mounted vertically on the tail boom. Sometimes called a Tail.

Variable Pitch—A type of wind turbine rotor where the attack angle of the blades can be adjusted either automatically or manually.

VAWT—See Vertical Axis Wind Turbine.

Vertical Axis Wind Turbine (VAWT)—A wind generator design where the rotating shaft is perpendicular to the ground, and the cups or blades rotate parallel to the ground.

Voltage—A measure of electromotive force. One Volt is the potential difference needed in a circuit to make one Ampere flow, dissipating one Watt of heat.

Watt—A unit of power that equals one Joule of electrical energy per second.

Weibull Distribution—A probability distribution, used to show how likely the wind is to be blowing at any given speed. See also Rayleigh Distribution.

Whiskey—Magical fluid consumed only well after all work is done. See Maker's Mark.

Wild AC—Alternating Current that varies in Frequency.

Wind Generator—A device that converts kinetic energy in the wind into electricity energy.

Windmill—A device that uses wind power to mill grain into flour. But informally used as a synonym for wind generator or wind turbine, and to describe machines that pump water with wind power. The authors are frequently guilty of all of these infractions.

Wind Turbine—See Wind Generator.

Windward—Toward the direction from which the wind blows.

Yaw—Rotation parallel to the ground. A wind generator Yaws to face winds coming from different directions.

Yaw Axis—Vertical axis through the center of gravity. The center of the top of the tower is on the Yaw Axis.

Zymurgy—The study of fermented, alcoholic beverages.

Appendix A — Tap drill size chart

Tapping threads into a hole is not quite as simple as it looks. As we discussed in the *Tips for tapping* sidebar in Chapter 8, *Coil winder*, if you want a threaded hole to accept a 5/16 inch bolt (which is actually properly called a "machine screw"), you don't use a 5/16 inch drill bit to bore the hole! Instead, you need to look up the bolt size and threads per inch (TPI) in the chart below, and find the proper drill bit size. Refer to that sidebar for more information on proper tapping technique.

Screw size (inches)	Screw TPI	Screw size, decimal inches	Drill bit size (inches, #, or letter)	Drill bit size, decimal inches
1/8	32	.0844	3/32	.0937
	40	.0925	38	.1015
9/64	40	.1081	32	.1160
5/32	32	.1157	1/8	.1250
	36	.1202	30	.1285
11/64	32	.1313	9/64	.1406
3/16	24	.1334	26	.1470
	32	.1467	22	.1570
13/64	24	.1490	20	.1610
7/32	24	.1646	16	.1770
	32	.1782	12	.1890
15/64	24	.1806	10	.1935
1/4	20	.1850	7	.2010
	24	.1959	4	.2090
	27	.2019	3	.2130
	28	.2036	3	.2130
	32	.2094	7/32	.2187
5/16	18	.2403	F	.2570
	20	.2476	17/64	.2656
	24	.2584	I	.2720

Screw size (inches)	Screw TPI	Screw size, decimal inches	Drill bit size (inches, #, or letter)	Drill bit size, decimal inches
	27	.2644	J	.2770
	32	.2719	9/32	.2812
3/8	16	.2938	5/16	.3125
	20	.3100	21/64	.3281
	24	.3209	Q	.3320
	27	.3269	R	.3390
7/16	14	.3447	U	.3680
	20	.3726	25/64	.3906
	24	.3834	X	.3970
	27	.3894	Y	.4040
1/2	12	.3918	27/64	.4219
	13	.4001	27/64	.4219
	20	.4351	29/64	.4531
	24	.4459	29/64	.4531
	27	.4519	15/32	.4687
9/16	12	.4542	31/64	.4844
	18	.4903	33/64	.5156
	27	.5144	17/32	.5312
5/8	11	.5069	17/32	.5312
	12	.5168	35/64	.5469
	18	.5528	37/64	.5781
	27	.5769	19/32	.5937
11/16	11	.5694	19/32	.5937
	16	.6063	5/8	.6250
3/4	10	.6201	21/32	.6562
	12	.6418	43/64	.6719
	16	.6688	11/16	.6875
	27	.7019	23/32	.7187

Appendix B — Wire gauge chart

American Wire Gauge (AWG) for copper wire

AWG = American Wire Gauge size from 0000 to 40

Ø-mils = Diameter in mils (1 mil = .001 inch)

Ø-mm = Diameter in millimeters.

CM = Cross sectional Area in Circular Mils. (circular mils = diameter in mils squared)

Ω/Kft = Ohms Per 1,000 ft.

Ft/Ω = Number of feet required for 1 Ohm of resistance

Ft/Lb = Feet Per Pound

Ω/Lb = Ohms Per Pound

Lb/Kft = Pounds Per 1,000 feet

AWG	Ø-mils	Ø-mm	CM	Ω/Kft	Ft/Ω	Ft/Lb	Ω/Lb	Lb/Kft
0000	459.99	11.684	211592	0.0490	20402	1.5613	0.0001	640.48
000	409.63	10.405	167800	0.0618	16180	1.9688	0.0001	507.93
00	364.79	9.2657	133072	0.0779	12831	2.4826	0.0002	402.80
0	324.85	8.2513	105531	0.0983	10175	3.1305	0.0003	319.44
1	289.29	7.3480	83690	0.1239	8069.5	3.9475	0.0005	253.33
2	257.62	6.5436	66369	0.1563	6399.4	4.9777	0.0008	200.90
3	229.42	5.8272	52633	0.1970	5075.0	6.2767	0.0012	159.32
4	204.30	5.1893	41740	0.2485	4024.7	7.9148	0.0020	126.35
5	181.94	4.6212	33101	0.3133	3191.7	9.9804	0.0031	100.20
6	162.02	4.1153	26251	0.3951	2531.1	12.585	0.0050	79.460
7	144.28	3.6648	20818	0.4982	2007.3	15.869	0.0079	63.014
8	128.49	3.2636	16509	0.6282	1591.8	20.011	0.0126	49.973
9	114.42	2.9063	13092	0.7921	1262.4	25.233	0.0200	39.630
10	101.90	2.5881	10383	0.9989	1001.1	31.819	0.0318	31.428
11	90.741	2.3048	8233.9	1.2596	793.93	40.122	0.0505	24.924
12	80.807	2.0525	6529.8	1.5883	629.61	50.593	0.0804	19.765
13	71.961	1.8278	5178.3	2.0028	499.31	63.797	0.1278	15.675
14	64.083	1.6277	4106.6	2.5255	395.97	80.447	0.2031	12.431
15	57.067	1.4495	3256.7	3.1845	314.02	101.44	0.3230	9.8579
16	50.820	1.2908	2582.7	4.0156	249.03	127.91	0.5136	7.8177
17	45.257	1.1495	2048.2	5.0636	197.49	161.30	0.8167	6.1997
18	40.302	1.0237	1624.3	6.3851	156.62	203.39	1.2986	4.9166

American Wire Gauge (AWG) for copper wire (continued)

AWG	Ø-mils	Ø-mm	CM	Ω/Kft	Ft/Ω	Ft/Lb	Ω/Lb	Lb/Kft
19	35.890	0.9116	1288.1	8.0514	124.20	256.47	2.0648	3.8991
20	31.961	0.8118	1021.5	10.153	98.496	323.41	3.2832	3.0921
21	28.462	0.7229	810.10	12.802	78.111	407.81	5.2205	2.4521
22	25.346	0.6438	642.44	16.143	61.945	514.23	8.3009	1.9446
23	22.572	0.5733	509.48	20.356	49.125	648.44	13.199	1.5422
24	20.101	0.5106	404.03	25.669	38.958	817.66	20.987	1.2230
25	17.900	0.4547	320.41	32.368	30.895	1031.1	33.371	0.9699
26	15.940	0.4049	254.10	40.815	24.501	1300.1	53.061	0.7692
27	14.195	0.3606	201.51	51.467	19.430	1639.4	84.371	0.6100
28	12.641	0.3211	159.80	64.898	15.409	2067.3	134.15	0.4837
29	11.257	0.2859	126.73	81.835	12.220	2606.8	213.31	0.3836
30	10.025	0.2546	100.50	103.19	9.6906	3287.1	339.18	0.3042
31	8.9276	0.2268	79.702	130.12	7.6850	4145.0	539.32	0.2413
32	7.9503	0.2019	63.207	164.08	6.0945	5226.7	857.55	0.1913
33	7.0799	0.1798	50.125	206.90	4.8332	6590.8	1363.6	0.1517
34	6.3048	0.1601	39.751	260.90	3.8329	8310.8	2168.1	0.1203
35	5.6146	0.1426	31.524	328.99	3.0396	10480	3447.5	0.0954
36	5.0000	0.1270	25.000	414.85	2.4105	13215	5481.7	0.0757
37	4.4526	0.1131	19.826	523.11	1.9116	16663	8716.2	0.0600
38	3.9652	0.1007	15.723	659.63	1.5160	21012	13859	0.0476
39	3.5311	0.0897	12.469	831.78	1.2022	26496	22037	0.0377
40	3.1445	0.0799	9.8880	1048.9	0.9534	33410	35040	0.0299

Compiled by Fr. Thomas McGahee in April 1998 based on information derived from various sources, and used here by permission.

Since there were some discrepancies between data sources, author McGahee chose not to just copy the numbers found in those sources, but rather to write a computer program that would generate the data based on a Best Fit mathematical model. The author checked the diameter in mils results against the data contained in "Machinery's Handbook," Twenty-First Edition, 1982, published by Industrial Press Inc.

All numbers shown are accurate to +/- 1 in the least significant digit position.

Appendix C — Metric conversions

During the entire long process of writing this book, the authors puzzled about how to provide construction details and diagrams that could be used worldwide. Unfortunately for this goal of ours, the United States, Liberia and Myanmar are the only countries in the world that have not yet officially adopted the metric (SI) system. That means that every nut, bolt, magnet, spool of wire, stick of pipe and flat of steel that we obtain locally is sized in inches and feet. We simply don't have a local source for components in metric sizes.

Some mixing of standard and metric parts would be fine for building the wind turbine in this book, but other combinations could be disastrous. For example, the magnet and wire sizes specified here must be followed exactly—larger or smaller sizes that seem "close enough" could drastically change alternator performance. On the other hand, substituting metric nuts, bolts and washers would be easy, and cause no problems at all. Somewhere in between would be substituting a metric bearing and spindle—everything would work the same, but all of the other frame parts would need their dimensions and cuts customized.

What we ended up with is this Appendix, with metric conversion formulas. Just remember not to mess with the magnet size or wire size, and you can probably adapt this wind turbine design to metric sizes with not much difficulty!

US to metric conversions

Symbol	To convert from	Multiply by	To get	Symbol
Length				
in	inches	25.4	millimeters	mm
ft	feet	0.305	meters	m
yd	yards	0.914	meters	m
mi	miles	1.61	kilometers	km
Area				
in²	square inches	645.2	sq. millimeters	mm²
ft²	square feet	0.093	square meters	m²
yd²	square yard	0.836	square meters	m²

Symbol	To convert from	Multiply by	To get	Symbol
Area (continued)				
ac	acres	0.405	hectares	ha
mi²	square miles	2.59	sq. kilometers	km²
Volume				
fl oz	fluid ounces	29.57	milliliters	ml
gal	gallons	3.785	liters	l
ft³	cubic feet	0.028	cubic meters	m³
yd³	cubic yards	0.765	cubic meters	m³
Mass				
oz	ounces	28.35	grams	g
lb	pounds	0.454	kilograms	kg
T	short tons	0.907	metric tons	t
Force				
lbf	foot-pounds	4.45	newtons	N
lbf/in²	foot-pounds per square inch	6.89	kilopascals	kPa

Metric to US conversions

Symbol	To convert from	Multiply by	To get	Symbol
Length				
mm	millimeters	0.039	inches	in
m	meters	3.28	feet	ft
m	meters	1.09	yards	yd
km	kilometers	0.621	miles	mi
Area				
mm²	millimeters	0.0016	square inches	in²
m²	square meters	10.764	square feet	ft²
m²	square meters	1.195	square yards	yd²
ha	hectares	2.47	acres	ac
km²	sq. kilometers	0.386	square miles	mi²
Volume				
ml	milliliters	0.034	fluid ounces	fl oz
l	liters	0.264	gallons	gal
m³	cubic meters	35.314	cubic feet	ft³
m³	cubic meters	1.307	cubic yards	yd³

Symbol	To convert from	Multiply by	To get	Symbol
Mass				
g	grams	0.035	ounces	oz
kg	kilograms	2.202	pounds	lb
t	metric tons	1.103	short tons	T
Force				
N	newtons	0.225	foot-pounds	lbf
kPa	kilopascals	0.145	foot-pounds per square inch	lbf/in²

Other metric conversions

One dozen (12) roses (US) = 10.12 roses (metric)

10^{12} microphones = 1 megaphone

10^{-6} phones = 1 microphone

2000 mockingbirds = 2 kilomockingbirds

10 cards = 1 decacards

10^{-6} fish = 1 microfiche

454 graham crackers = 1 pound cake

10^{-12} pins = 1 terrapin

10^{21} picolos = 10^{9} los = 1 gigolo

10 rations = 1 decoration

10 millipedes = 1 centipede

3-1/3 tridents = 1 decadent

10 monologs = 5 dialogues

5 dialogues = 1 decalogue

2 monograms = 1 diagram

8 nickels = 2 paradigms

2 snake eyes = 1 paradise

2 wharves = 1 paradox

10^{-2} mental = 1 centimental

10^{12} bulls = 1 terabull

10^{-12} boos = 1 picoboo

Appendix D — Useful energy data

A ctually, only a few of these are really useful at all. But many of them are quite interesting! All of the information in this appendix is from a very interesting and little known government agency—the US Energy Information Administration (EIA, *www.eia.doe.gov*). The busy statisticians there compile an amazing array of information. For example, are you curious how the price of electricity compares between the US, Canada, Europe and Asia? The information is at your fingertips (and updated regularly), along with data on how much the average consumer in each country uses per month. Our thanks also to the Wisconsin K-12 Energy Education Program (KEEP), we copied some of their ideas about how to organize all of this information.

Power conversions

	foot-pounds per second	horsepower	calories per second	kilowatts	watts
1 foot-pound per second	1	0.001818	0.3238	0.001356	1.356
1 horsepower	550	1	178.1	0.746	746
1 calorie per second	3.088	0.005615	1	0.004187	4.187
1 kilowatt	737.6	1.341	238.8	1	1000
1 watt	0.7376	0.001341	0.2388	0.001	1

Energy conversions

	British thermal unit (BTU)	foot-pounds	joules	calories	kilo-calories	kilowatt-hours
1 British thermal unit (BTU)	1	777.9	1055	252.0	0.252	0.000293
1 foot-pound	0.001285	1	1.356	0.3238	0.000323	0.000000377
1 joule	0.0009481	0.7376	1	0.2388	0.0002388	0.000000278
1 calorie	0.003969	3.088	4.187	1	0.001	0.000001163
1 kilocalorie	3.969	3088	4187	1000	1	0.001163
1 kilowatt hour	3413	2655000	3600000	859800	859.8	1

Average energy content of fuels

1 kilowatt hour of electricity	3413 British thermal units (BTU)
1 cubic foot of natural gas	1,008 to 1,034 BTU
1 therm of natural gas	100,000 BTU
1 gallon of crude oil	138,095 BTU
1 barrel of crude oil	5,800,000 BTU
1 gallon of residual fuel oil	149,690 BTU
1 gallon of gasoline	125,000 BTU
1 gallon of ethanol	84,400 BTU
1 gallon of methanol	62,800 BTU

Average energy content of fuels (continued)

1 gallon of gasohol (10% ethanol, 90% gasoline)	120,900 BTU
1 gallon of E-85 (85% ethanol, 15% gasoline)	90,500 BTU
1 gallon of kerosene or light distillate oil	135,000 BTU
1 gallon of middle distillate or diesel fuel oil	138,690 BTU
1 gallon of liquefied petroleum gas (LPG)	95,475 BTU
1 pound of coal	8,100 to 13,000 BTU
1 ton of coal	16,200,000 to 26,000,000 BTU
1 ton of wood	9,000,000 to 17,000,000 BTU
1 standard cord of wood	18,000,000 to 24,000,000 BTU
1 face cord of wood	6,000,000 to 8,000,000 BTU
1 pound of low pressure steam (recoverable heat)	1,000 BTU

US residential energy consumption by end use

Source: US Energy Information Administration

End use	Survey year				
	1987	1990	1993	1997	2001
Air-conditioning	15.8	15.9	13.9	11.8	16.0
Space heating	10.3	10.0	12.4	11.4	10.1
Water heating	11.4	11.2	10.3	11.0	9.1
Total appliances	62.5	63.0	63.4	65.9	64.7

Average US household electric consumption for major electrical appliances, 1998

Source: US Energy Information Administration

Appliance	Annual kwh consumption per household
Central air conditioning	2,667
Room air conditioning	738
Water heater	2,671
Refrigerator	1,155
Freezer	1,204
Range / oven	458
Dishwasher	299
Water bed heater	960
Clothes washer	99
Clothes dryer	875

Energy-efficient, multi-use appliances

Kodiak the dog says that *she* is truly the best, most efficient multipurpose appliance ever invented. She saves hot water by prewashing dishes, propane and firewood by helping warm the bed, and electricity by vacuuming food scraps off the floor. She controls destructive and evil squirrels without traps or poisons, and also serves as a non-electric home security system to warn of and scare off both 4-legged and 2-legged intruders. All for the low energy consumption of 3 cups of kibbles per day!

Appendix E — Tools

Once again, the authors puzzled for a long time over this Appendix about tools. There are so many different ways you can complete each step in the wind turbine construction process, it would be impossible to list them. In fact, much of the time the construction processes we've developed here at Otherpower HQ were based simply on which tools we had available!

Look at it this way—let's say you have a nut that you need to remove from a bolt. The best tool for the job would be a socket and ratchet, but perhaps the end of the bolt is too long and you don't have a deep enough socket. That would call for an end wrench of the proper size. Then let's say that wrench is not available, and you are in the wilds of Alaska building your turbine with no hardware store nearby. An adjustable (Crescent®) wrench would certainly do the trick, but your dog grabbed it and hid it in the woods with his pile

Figure E.1—I just *know* that wrench is in here somewhere!

of squeaky toys and rawhide chewies. A pair of Vise-Grips® would be the next best choice, but the tool shed is such a mess you can't find them, even after looking for two hours (Figure E.1, the owner of which shall remain anonymous). All you can find is a rusty old pair of pliers by the compost heap. Not the ideal tool for the job, but will likely work just fine—and sure beats a long trip to town to buy a wrench! Where the authors live, a town trip costs nearly $20 in gasoline alone, which is more than what the new wrench would cost.

With that in mind, here's a list of tools that we use at our shop for building wind turbine kits and parts. Many of them are completely redundant—we don't use all these tools to build a wind turbine every time, but we have used all of them for different things in the past. Don't head to the hardware store and buy every tool listed below, though that would surely make your local hardware dealer very, very happy! Instead, get a feel for how each process in the construction is going, and buy tools for the steps that are causing you a royal pain in the butt.

Another issue is if you have electric power available at your workshop or not. It's entirely possible to build this entire wind turbine using hand tools only—nothing powered by electricity is actually *required*. That could be very important in a remote area where there is no electricity—which is most likely where this wind turbine would be needed most. However, electric hand tools will save you many hours and days of construction time! And fixed-base electric shop tools will give your project a much higher level of precision. If you buy any components of the wind turbine as pre-fabricated from us, the need for some of these tools may be eliminated—you'll need to check each fabrication chapter in advance to see which steps have already been done for you.

Measurement tools

You'll need to make numerous accurate measurements during every step of constructing the wind turbine, and you'll be making marks on wood, metal, and fiberglass. Accurate measurements are a must!

• Tape measure

• Steel ruler

• Set square

• Angle/bevel gage

• Protractor

• Scribe

• Compass

• Spirit level

• Vernier or dial calipers

• Pencils

• Permanent markers

• Soapstone welding marker

Safety equipment

Chapter 5, *Shop safety*, covers the numerous safety precautions you'll have to keep in mind for all steps of turbine construction. The list below summarizes the safety gear you'll need:

• Safety glasses or goggles

• Hearing protection

• Dust mask

• Respirator for resin

• Leather gloves

• Work boots

• Welding mask

• Welding apron

• Latex gloves

Electrical tools

- True RMS multimeter
- Laser tachometer
- Alligator clip jumper wires
- Large soldering iron
- Propane torch
- Rosin core electrical solder
- Heat shrink tubing
- Wire cutters

Woodworking tools

- Workbench
- Electric hand drill
- Drill bits
- Drill press
- Brace and bit
- Hole saw set
- Cordless drill/screwdriver
- Electric belt sander
- Electric finish sander
- Bench belt/disc sander
- Sandpaper
- Bandsaw
- Electric jigsaw
- Hand saw
- Drawknife
- Spoke shave
- Electric planer

- Hand planes, large and small
- Wood chisel set
- Calipers and dividers
- Bench vise
- C clamps
- Hammer
- Wooden mallet
- Rubber mallet
- Sharpening stones

Metalworking tools

- Metal workbench
- Electric drill
- Drill press
- Metal cutting bandsaw (bench)
- Metal cutting bandsaw (portable)
- Chop saw
- Hacksaw
- Gas cutting torch and accessories
- Welder, 200 amp wire feed, and accessories
- Hole saws, high-quality bi-metal
- Vise
- C clamps and other welding clamps
- Bench grinder
- Angle grinder
- Metal files
- Taps and dies

- Center punch
- Screwdrivers, flat and phillips head
- Adjustable wrench
- End wrench set
- Socket wrenches and ratchet

Casting tools

- Plastic mixing bowls
- Plastic mixing spoons
- Utility knife
- Drop cloth
- Shop rags
- Furniture polish (mold release compound)

Miscellaneous consumables

- Welding rod
- Saw blades
- Grinding wheels and disks
- Sandpaper
- Cyanoacrylate glue (Superglue®)
- Cyanoacrylate glue accelerator
- Thread locking compound
- Axle grease
- Lead sheet (or duck decoy weights)
- Wood screws

- Extra magnets (handy in the shop)
- Wood stain
- Linseed oil
- Wood putty for minor repairs
- Primer and paint for metal
- Duct tape
- Electrical tape

Borzoi Haiku
Right tool for the job
My long nose is prehensile
I do not need thumbs

Appendix F — Useful wind data

Power output per swept area

Wind power researcher Mike Klemen came up with the theoretical concepts of the "perfect wind turbine" and the "good wind turbine" for one of his articles. We discuss these concepts in Chapter 3, *Power in the wind,* but in brief: A "perfect" wind turbine performs with a cP right at the Betz Limit of 0.5926 at all wind speeds—and no such turbine exists. A "good" wind turbine performs with a cP of 0.35 at all wind speeds—and while some commercial small turbines may perform at this cP in certain wind speeds, none will do it through the entire range. The figures shown are per square foot (ft²) and square meter (m²) of swept area.

	Power output (watts)			
Wind speed	**Perfect turbine**	**Good turbine**	**Perfect turbine**	**Good turbine**
mph [m/s]	per m²	per m²	per ft²	per ft²
1 - [0.45]	0.031	0.019	0.00295	0.00175
2 - [0.89]	0.254	0.151	0.0236	0.0140
3 - [1.34]	0.857	0.508	0.0796	0.0472
4 - [1.79]	2.031	1.205	0.1887	0.1119
5 - [2.24]	3.966	2.353	0.3685	0.2186
6 - [2.68]	6.854	4.066	0.6367	0.3777
7 - [3.13]	10.88	6.457	1.01	0.5998
8 - [3.58]	16.25	9.638	1.51	0.8954
9 - [4.02]	23.13	13.72	2.15	1.28
10 - [4.47]	31.73	18.82	2.95	1.75
11 - [4.92]	42.23	25.05	3.92	2.33
12 - [5.36]	54.83	32.53	5.09	3.02
13 - [5.81]	69.71	41.36	6.48	3.84
14 - [6.26]	87.07	51.65	8.09	4.80
15 - [6.71]	107.1	63.53	9.95	5.90
16 - [7.15]	130.0	77.10	12.07	7.16
17 - [7.60]	155.9	92.48	14.48	8.59
18 - [8.05]	185.1	109.8	17.19	10.2
19 - [8.49]	217.6	129.1	20.22	12.0
20 - [8.94]	253.9	150.6	23.58	14.0
21 - [9.39]	293.9	174.3	27.30	16.2
22 - [9.83]	337.9	200.4	31.39	18.6
23 - [10.28]	386.1	229.0	35.87	21.3
24 - [10.73]	438.7	260.2	40.75	24.2
25 - [11.18]	495.8	294.1	46.06	27.3
26 - [11.62]	557.7	330.8	51.81	30.7

Energy output per month per swept area

As we discussed in Chapter 3, *Power in the wind*, energy output per month from your wind turbine is much more important to your renewable energy system than the instantaneous power output of a turbine. You can have the fanciest and most expensive turbine in the world up and flying, but if it is mounted on a low tower, is surrounded by obstructions or is in an area with poor average wind speeds, monthly power output will likely be dismal. The chart below, also from Mike Klemen, shows the predicted energy output in kilowatt-hours (kwh) per month for turbines at different average wind speeds. Like the previous chart, it's calculated per square foot (ft²) and square meter (m²) of swept area and covers both the theoretical "perfect" (cP=0.5926) and "good" (cP=0.35) turbines. Thanks again, Mike!

Energy output per month (kwh)

Average wind speed mph - [m/s]	Perfect turbine per m²	Good turbine per m²	Perfect turbine per ft²	Good turbine per ft²
5 - [2.24]	4.47	2.65	0.415	0.246
6 - [2.68]	7.99	4.74	0.742	0.440
7 - [3.13]	12.98	7.70	1.206	0.715
8 - [3.58]	19.66	11.66	1.826	1.083
9 - [4.02]	28.02	16.63	2.604	1.545
10 - [4.47]	37.70	22.36	3.502	2.078
11 - [4.92]	47.95	28.45	4.455	2.643
12 - [5.36]	57.96	34.38	5.384	3.194
13 - [5.81]	67.01	39.75	6.226	3.693
14 - [6.26]	74.68	44.30	6.938	4.116

Wind turbulence from a ground-level obstruction

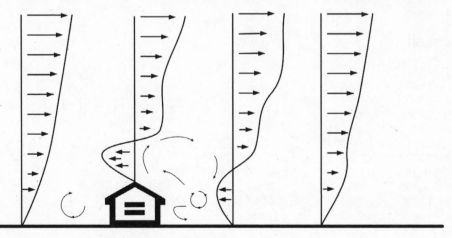

*Source: Adapted from the book **Wind Energy** by Kovarik, Pipher and Hurst.*

Theoretical increase in power available in the wind by tower height

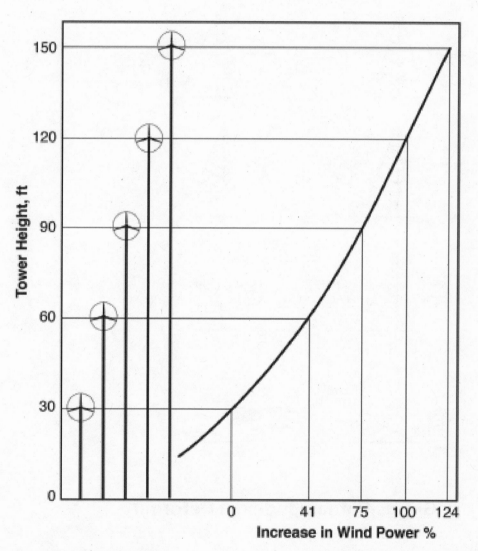

Source: US Department of Energy

Estimating average wind speed by tree flagging

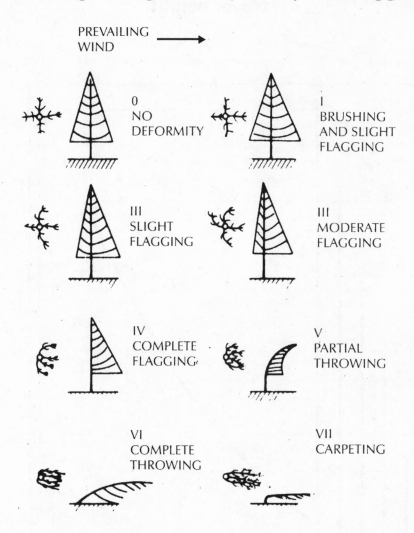

PREVAILING WIND →

0 NO DEFORMITY

I BRUSHING AND SLIGHT FLAGGING

III SLIGHT FLAGGING

III MODERATE FLAGGING

IV COMPLETE FLAGGING

V PARTIAL THROWING

VI COMPLETE THROWING

VII CARPETING

Griggs-Putnam Index of Deformity

Average wind speed	Index						
	I	II	III	IV	V	VI	VII
mph	7-9	9-11	11-13	13-16	15-18	16-21	22+
m/s	3-4	4-5	5-6	6-7	7-8	8-9	10

Source: Battelle PNL

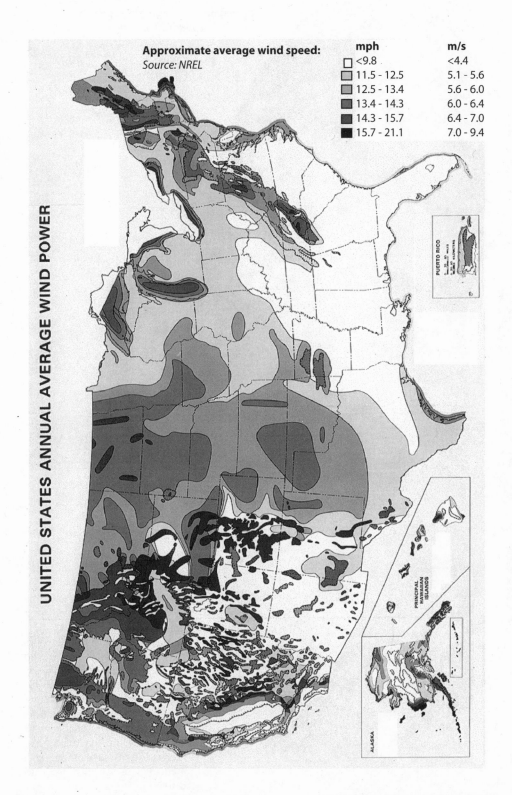

Approximate average wind speed:
Source: NREL

	mph	m/s
	<9.8	<4.4
	11.5 - 12.5	5.1 - 5.6
	12.5 - 13.4	5.6 - 6.0
	13.4 - 14.3	6.0 - 6.4
	14.3 - 15.7	6.4 - 7.0
	15.7 - 21.1	7.0 - 9.4

UNITED STATES ANNUAL AVERAGE WIND POWER

PUERTO RICO

PRINCIPAL HAWAIIAN ISLANDS

ALASKA

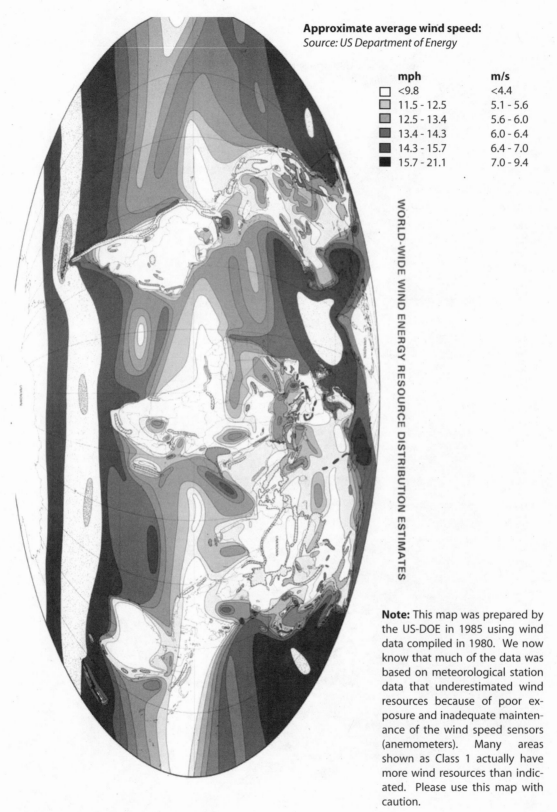

Approximate average wind speed:
Source: US Department of Energy

	mph	m/s
☐	<9.8	<4.4
	11.5 - 12.5	5.1 - 5.6
	12.5 - 13.4	5.6 - 6.0
	13.4 - 14.3	6.0 - 6.4
	14.3 - 15.7	6.4 - 7.0
■	15.7 - 21.1	7.0 - 9.4

WORLD-WIDE WIND ENERGY RESOURCE DISTRIBUTION ESTIMATES

Note: This map was prepared by the US-DOE in 1985 using wind data compiled in 1980. We now know that much of the data was based on meteorological station data that underestimated wind resources because of poor exposure and inadequate maintenance of the wind speed sensors (anemometers). Many areas shown as Class 1 actually have more wind resources than indicated. Please use this map with caution.

Index